ABOUT THE AUTHOR

From the day he was found in a carrier bag on the steps of Guy's Hospital in London, **Andy McNab** has led an extraordinary life.

As a teenage delinquent, Andy McNab kicked against society. As a young soldier, he waged war against the IRA in the streets and fields of South Armagh. As a member of 22 SAS, he was at the centre of covert operations for nine years, on five continents. During the Gulf War he commanded Bravo Two Zero, a patrol that, in the words of his commanding officer, 'will remain in regimental history for ever'. Awarded both the Distinguished Conduct Medal (DCM) and Military Medal (MM) during his military career, McNab was the British Army's most highly decorated serving soldier when he finally left the SAS.

Since then Andy McNab has become one of the world's best-selling writers, drawing on his insider knowledge and experience. As well as three non-fiction bestsellers – including *Bravo Two Zero*, the bestselling British work of military history – he is the author of the Nick Stone and Tom Buckingham thrillers. He has also written a number of books for children.

Besides his writing work, he lectures to security and intelligence agencies in both the USA and UK, works in the film industry advising Hollywood on everything from covert procedure to training civilian actors to act like soldiers, and he continues to be a spokesperson and fundraiser for both military and literacy charities.

WOMEN IN SCIENCE AND ENGINEERING:

Increasing Their Numbers in the 1990s

A Statement on Policy and Strategy

from the

Committee on Women in Science and Engineering

**Office of Scientific and Engineering Personnel
National Research Council**

NATIONAL ACADEMY PRESS
Washington, D.C.   1991

National Academy Press • 2101 Constitution Avenue, NW • Washington, DC 20418

COMMITTEE ON WOMEN
IN SCIENCE AND ENGINEERING

*MILDRED S. DRESSELHAUS, Massachusetts Institute of Technology, *Chair*

BETSY ANCKER-JOHNSON, General Motors Corporation

GEORGE CAMPBELL, JR., National Action Council for Minorities in Engineering

*JEWEL PLUMMER COBB, California State University—Los Angeles

*ESTHER M. CONWELL, Xerox Corporation

BRUCE ANDREW FOWLER, University of Maryland Medical School

LILLI S. HORNIG, Wellesley College

PAT HILL HUBBARD, American Electronics Association

SHIRLEY A. JACKSON, AT&T Bell Laboratories

*THOMAS E. MALONE, Association of American Medical Colleges

CORA B. MARRETT, University of Wisconsin—Madison

*MARSHA LAKES MATYAS, American Association for the Advancement of Science

GIAN-CARLO ROTA, Massachusetts Institute of Technology

*GARRISON SPOSITO, University of California—Berkeley

KAREN K. UHLENBECK, University of Texas—Austin

Staff Officer: Linda Dix
Senior Secretary: Gaelyn Davidson

* Members of the Panel on Strategic Planning

FOREWORD

From time to time, it is necessary to alert the research and policy communities to opportunities for action in areas of mutual concern. One such area is the participation and utilization of women in science and engineering in the United States. The underparticipation of women in these fields is a matter of record; the understanding that something might be done to bring qualified women into productive careers as researchers, teachers, and practitioners of science and engineering prompted the National Research Council (NRC) in 1990 to establish the Committee on Women in Science and Engineering (CWSE) within the Office of Scientific and Engineering Personnel (OSEP). The Committee met for the first time in 1991 and quickly developed a plan of action—a strategic plan—for effecting change. That plan is the subject of this report.

In developing its strategic plan, the Committee on Women in Science and Engineering began its work by surveying the global policy environment, including policies affecting the employment and education of women in science and engineering. Rates of participation of women in science and engineering have been found to be affected significantly by economic globalization and demographic factors. That is, employers confronting dramatic shifts in the comparative advantage of U.S. industries in an international marketplace are increasingly motivated to maximize the productivity of an increasingly diverse R&D work force. For women with an interest in industry-based careers, the growing emphasis on "diversity for economic productivity" may actually create a positive climate for their employment—representing a significant change from earlier conditions. Whereas demographic trends have created more opportunities for women in industry, fewer positive changes are evident in the employment of female scientists and engineers in academe and in the public (government) sector. Rates of advancement continue to be quite low compared to the advancement of men, although the reasons for the differences are complex and diverse.

The Committee on Women in Science and Engineering tackles these issues head on. They do so by exploring the capacity of the education infrastructure to prepare women for careers in science and engineering and follow that discussion by exploring the factors that determine how a skilled female scientist or engineer pursues a career—the comparative role of postdoctoral training and the issues surrounding the advancement of women in various employment sectors.

A wide spectrum of programs has been introduced to assist women to gain entry into science and engineering through intervention efforts at various stages of education and employment. As the Committee points out, however, the absence of systematic and reliable information on the effects of these interventions represents a major barrier to the evaluation of policies and intervention programs fostering careers for women in science and engineering. Thus, the Committee has begun to shape its plan of action based on a call for increased information and more rigorous analysis.

The success of any strategic plan will depend on at least two factors: a clear statement of the program goals and the specification of an effective plan to achieve those goals. The Committee on Women in Science and Engineering has taken an important first step by outlining its strategy for the foreseeable future, for which they are commended. Because the Committee does not act alone, but rather responds to the concerns of its various sponsors and other partners in this effort, we can look forward to an increasingly rich program of activities from the Committee in coming years as it establishes itself firmly in the research and policy world.

> *Linda Wilson*
> Chair
> OSEP Advisory Committee
> on Studies and Analyses

PREFACE AND ACKNOWLEDGMENTS

The Committee on Women in Science and Engineering (CWSE), a continuing committee established within the National Research Council (NRC) in January 1991, has a four-pronged mandate to:

1. collect and disseminate current data about the participation of women in science and engineering to broad constituencies in academe, government, industry, and professional societies;
2. monitor the progress of efforts to increase the participation of women in scientific and engineering careers;
3. conduct symposia, workshops, and other meetings of experts to explore the policy environment, to stimulate and encourage initiatives in program development for women in science and engineering, and to evaluate their effectiveness on a regular basis; and
4. propose research and conduct special studies on issues particularly relevant to women scientists and engineers in order to develop reports that will document evidence and articulate NRC recommendations for action.

CWSE activities will focus on the participation of women in science and engineering at the postsecondary level of education and in various employment sectors.

This report is the culmination of an initial examination by CWSE of the status of women scientists and engineers in the United States. In addition to providing statistics on the participation of women in the education/ employment pipeline, it summarizes the Committee's deliberations relating to its role in increasing the participation and improving the status of women in science and engineering.The report further offers an ambitious strategic plan of both short-term and long-term activities. As evidenced by the comprehensiveness of this first report, issued within its first year of operation, the Committee on Women in Science and Engineering has taken a proactive stance in this arena. By stimulating research on issues relevant to women scientists and engineers, by establishing study panels that can explore some subset of these issues in greater depth, and by briefing appropriate officials on matters leading to the development of programs for women in science and engineering, this Committee plans to keep the issue of women's participation in science and engineering at the forefront.

The Committee on Women in Science and Engineering appreciates the assistance that it received from a number of individuals. The earlier Planning Group on Possible OSEP Initiatives for Increasing the Participation of Women in Scientific and Engineering Careers laid the groundwork for this Committee: William O. Baker, Jewel Plummer Cobb, Edward E. David, Mildred S. Dresselhaus, Marsha Lakes Matyas, Karen K. Uhlenbeck, and Harriet Zuckerman. Valuable advice was received from OSEP's former chairman, William D. Carey, and from Esther M. Conwell, who served as the liaison to the Planning Group from OSEP's Advisory Committee on Studies and Analyses.

Initial financial support was provided by the National Academy of Engineering and the Andrew W. Mellon Foundation, through the efforts of Bruce Guile and Harriet Zuckerman, respectively. In addition, representatives from other sponsoring organizations have assisted the Committee in its development of the strategic plan found in Chapter 6 of this report: Charles R. Bowen, director of plans and program administration, Office of University Relations, IBM Corporation; Burt H. Colvin, director for academic affairs, National Institute of Standards and Technology, Department of Commerce; Margaret G. Finarelli, associate administrator for external affairs, and Frank Owens, manager of special projects in resources and management, National Aeronautics and Space Administration; Marguerite Hays, administrator for research, VA Medical Center-Palo Alto, Department of Veterans Affairs; Margrete S. Klein, director of women's programs, Division of Research Initiation and Improvement, National Science Foundation; Mark Myers, vice president for corporate research, Xerox Corporation; and Richard E. Stephens, director of university and industry programs, Office of Energy Research, Department of Energy.

We hope that the efforts of the many individuals involved in this exploratory examination of the current status of women in science and engineering in the United States will clarify the issues and assist policy makers, researchers, and institutions of both education and employment in developing their agenda for increasing the participation of women in these disciplines during the next decade.

Mildred S. Dresselhaus
Chairman

CONTENTS

Executive Summary   1

1   Introduction   5
    • The Global Policy Environment   7
    • Conclusions   26

2   The Science and Engineering Education Infrastructure   29
    • Formal Mechanisms   30
    • Informal Mechanisms   42
    • Priority Issues   48

3   Effectiveness of Intervention Models   51
    • Precollege Programs   52
    • Undergraduate Programs   53
    • Graduate Programs   57
    • Career Interventions   59
    • Priority Issues   68

4   Career Patterns   71
    • Postdoctoral Appointments   71
    • Employment in a Scientific or Engineering Field   76
    • Priority Issues   89

5   Measurement for Scientific and Engineering Human
    Resources   91
    • Education Infrastructure   91
    • Intervention Strategies   95
    • Career Patterns   98
    • Priority Issues   102

6   Strategic Plan for Increasing the Numbers of Women
    in Science and Engineering   105
    • First-Year Plan   116
    • Long-Range Plan   107

Bibliography   115
Related Tables   127
Technical Appendix   143

LIST OF TABLES

1: Civilian Employment of Scientists, Engineers, and Technicians (SET), by Field, 1986 and 2000 — 8

2: Bachelor's Degrees in Science and Engineering as Percentage of All Baccalaureates Awarded, Selected Years, 1972-1989 — 12

3: Ph.D.s Awarded by U.S. Universities to Non-U.S. Citizens, 1989 — 15

4: Science and Engineering Degrees Granted to Women, by Degree Level, 1986 and 1989 — 18

5: Degrees to Women in Physics and Women as Physics Faculty (in percent) — 21

6: Employers of Doctorate Recipients in Science and Engineering, by Sex, 1989 — 23

7: Academic Ranks of All U.S. Doctorate Recipients in Science and Engineering, 1989 — 25

8: College Major Field of Study of 1980 High School Seniors Who Had Graduated from College by 1986, by Intended Field of Study in High School and by Sex (in percent) — 31

9: 1980 High School Seniors Who Graduated from College by 1986, by Major Field of Study and by Race/Ethnicity and Sex — 32

10: Top 25 Science and Engineering Degree-Granting Institutions, 1980-1990 (all graduates) — 34

11: Top Five Baccalaureate Institutions of Female Science and Engineering Doctorate Recipients, by Field of Doctorate, 1985-1990 — 36

12: Percentage Distribution of Primary Sources of Support of Doctorate Recipients, by Sex and Broad Field, 1987 and 1989 — 38

13: NSF Graduate Fellowship Program Applications and Awards, by Sex, 1985 and 1990 — 40

14: NSF Minority Graduate Fellowship Program Applications and Awards by Sex, 1985 and 1990 — 44

15: Postgraduation Plans of Science and Engineering Doctorates (U.S. citizens only), 1985-1989 — 73

16: Reasons Given for Not Being Employed Full-Time by Science and Engineering Doctorate Recipients, 1989 — 78

17: Tenure Status of All U.S. Doctorate Recipients in Science and Engineering, 1989 — 81

18: Numbers of Women Employed in S&E Fields, by Race/Ethnicity: 1982-1986 — 96

A: Top 25 Science and Engineering Degree-Granting Institutions, 1980-1990 (men only)   128
B: NSF Graduate Fellowship Program Applications and Awards, by Sex, 1985-1991   130
C: NSF Graduate Fellowship Program Applications by Women, by Ethnic Group, 1989-1991   134
D: NSF Graduate Fellowship Program Awards to Women, by Ethnic Group, 1989-1991   136
E: NSF Minority Graduate Fellowship Program Applications by Women, by Ethnic Group, 1989-1991   138
F: NSF Minority Graduate Fellowship Program Awards to Women, by Ethnic Group, 1989-1991   140

LIST OF FIGURES

1. Science and engineering (S&E) pipeline, from high school through Ph.D. degree, **10**
2. U.S. population, aged 18-24, 1970-1988, and projected, 1990-2010 (in thousands), **11**
3. High and low estimates of the number of new Ph.D. faculty hires in the sciences and engineering, every five years, 1980-2015, **13**
4. U.S. and foreign engineering faculty, age 35 or less, 1973-1989, **14**
5. Percentage of women among employed scientists and engineers, by field, 1986, **20**
6. Women doctorates in science and engineering jobs, by field, 1973 and 1985, **22**
7. Women, blacks, and Hispanics in the federal work force, 1988 (in percent), **26**
8. S/E underemployment rates of men and women, by field, 1986, **77**
9. Average annual employment growth by sector of employment and sex, 1976-86, **83**
10. Average annual employment growth rate of scientists and engineers, by field and sex, 1978-1988, **84**

EXECUTIVE SUMMARY

For several decades, the National Research Council (NRC) has provided the federal government with information needed to develop effective policies for recruiting and retaining individuals in scientific and engineering (S&E) careers. Within the NRC, the Office of Scientific and Engineering Personnel (OSEP) is the focus for providing information and advice on the health of the human resource base. And the Committee on Women in Science and Engineering, established within OSEP in 1990, is charged with undertaking activities to facilitate the entry and retention of a greater number of talented women into careers in scientific and engineering disciplines.

The policy environment for the recruitment and retention of women in science and engineering can be characterized currently by attention to three types of issues: (1) demographic considerations, including the rising proportion of non-U.S. citizens in the U.S. work force; (2) education issues, with emphasis on the low rate of participation of women in the various fields of science and engineering; and (3) employment conditions in the U.S. work force. The decisions that we make about our S&E cadre today will have a significant effect on our ability to find solutions to future problems. Our ultimate success depends upon the degree to which we maximize use of all of the Nation's human resources.

Based on this policy environment, the Committee concludes that there is great potential for increasing the number of women in science and engineering, especially in areas where they are most underrepresented and where the national need is greatest. CWSE has therefore formulated a

1

plan of action on four topics. Following the Introduction, each of these topics is discussed in separate chapters within the report:

1. strengthening the **S&E education infrastructure**,
2. examining the effectiveness of **intervention programs** in sustaining the flow of women into science and engineering, and
3. exploring **career patterns** for women in S&E employment,
4. examining the adequacy of **data** available to measure the participation of women in science and engineering and the effectiveness of interventions designed to increase that participation.

S&E Education Infrastructure:
- identifying educational programs that have been effective in facilitating the recruitment and retention of women in S&E careers, with emphasis on programs at the undergraduate and graduate levels of education;

Intervention Strategies:
- encouraging the development of reliable outcome measures to assess the specific contribution of programs that enhance the flow of women into S&E careers;

Career Patterns:
- developing a program of studies to facilitate the positive employment opportunities related to diversification in the workplace; and
- exploring issues related to the support infrastructure that makes it

possible for women with family responsibilities to participate in the S&E labor force;

Measurement:

• fostering the development of finer measures of labor force adjustment, including tracking the career paths of postdoctoral personnel.

Each chapter concludes with a list of priority issues that the Committee believes warrant particular examination. Finally, in Chapter 6, the Committee delineates priorities for its short-term and long-range activites to increase the participation of women in science and engineering.

1

INTRODUCTION

The National Research Council (NRC) has a long and distinguished history of providing the federal government with information that is needed to develop effective policies for recruiting and retaining individuals in scientific and engineering (S&E) careers. In recent years, the Office of Scientific and Engineering Personnel (OSEP) has served as the focal point in NRC for providing information and advice on the health of the human resource base. Issues affecting women in science and engineering have been variously addressed over the last two decades (see Technical Appendix, which refers to earlier NRC efforts in this area). While some progress has been made in facilitating the entry of talented women into careers in these areas, much remains to be done in both recruiting and retaining women in science and engineering. It is no surprise, therefore, that to strengthen and clarify policies affecting the preparation and recruitment of women for careers in this area, the Governing Board of NRC concluded in 1988 that an ongoing effort was needed and requested OSEP to establish a committee that would have as its long-range goal the increased participation of women in the scientific and engineering work force.

The Committee on Women in Science and Engineering (CWSE) was established in 1990 and held its first meeting in March 1991. As a standing committee of NRC, CWSE includes in its growing portfolio four sets of activities:

1.   collecting and disseminating current data about the participation of

women in science and engineering to broad constituencies in academe, government, industry, and professional societies;

2. monitoring the progress of efforts to increase the participation of women in scientific and engineering careers;

3. conducting symposia, workshops, and other meetings of experts to explore the policy environment, to stimulate and encourage initiatives in program development for women in science and engineering, and to evaluate their effectiveness on a regular basis; and

4. proposing research and conducting special studies on issues particularly relevant to women scientists and engineers in order to develop reports that will document evidence and articulate NRC recommendations for action.

Specifically, CWSE will focus on the postsecondary segments of the education/employment pipeline—undergraduate, graduate, postdoctoral, and career segments—while keeping abreast of developments in precollege science education designed to recruit females into scientific and engineering careers.

The challenge in the 1990s will be to identify new opportunities for assuring that women will take their place beside men in building a strong science and technology base in the United States. This report outlines the role that the Committee on Women in Science and Engineering expects to take in achieving that goal.

The Global Policy Environment

Policies affecting the recruitment, education, and employment of women in science and engineering do not arise in a vacuum. Shifts in economic conditions, demographic patterns, and national research and development (R&D) goals stimulate the formulation of human resource policies and the selection of program goals (see, for example, Wildarsky, 1979; OSEP, 1991). The United States now faces a critical period in setting its technological and scientific priorities, and particular attention is being given to the expansion of the present pool of scientific and technical talent. It should come as no surprise that many policies affecting the role of women in science and engineering take as their starting point trends in the U.S. demography.

Demographic Issues

The Bureau of Labor Statistics predicts that the human-resource needs for science and engineering will increase, by 36 percent between the years 1986 and 2000, because of high-technology industrial growth and the increasing use of high-technology goods and services (see Table 1). How will we meet these increased human-resource needs? Richard C. Atkinson, chancellor of the University of California-San Diego and former director of the National Science Foundation (NSF), is emphatic about the current situation:

> Persuading more students to pursue graduate eduation in science and engineering, maintaining the vitality of our universities, raising the level of technological literacy, and making more effective use of the results and

7

TABLE 1: Civilian Employment of Scientists, Engineers, and Technicians (SET), by Field, 1986 and 2000

Field	Number Employed, 1986	Projected Percentage Increase in Employment, 2000
TOTAL, SET Fields	4,245,600	36
Total Scientists*	1,131,600	45
Computer Specialists	331,000	76
Life	140,000	21
Mathematical	48,000	29
Physical	180,000	13
Social	432,600	36
Total Engineers	1,371,000	32
Aeronautical/astronautical	53,000	11
Chemical	52,000	15
Civil	199,000	25
Electrical/electronics	401,000	48
Industrial	117,000	30
Mechanical	233,000	33
Other	316,000	24
Total Technicians	1,743,000	36
Computer Programmers	479,000	70
Draftsmen	348,000	2
Electrical/electronics	313,000	46
Other engineering	376,000	26
Physical, mathematical, and life sciences	227,000	15

* Includes 97,300 environmental scientists.
SOURCE: U.S. Department of Labor, Bureau of Labor Statistics, *Outlook 2000* (Bulletin 2302), Washington, D.C.: U.S. Government Printing Office, 1990.

insights of science in policy and decision making are not
separate problems. Rather, they are related components
of the fundamental question of the adequacy of our
science-education system and its relevance to the country's
needs (Atkinson, 1988).

After reviewing those needs, Eileen Collins concluded that

If present trends continue, there will be a shortage of
trained engineers which cannot be filled by the natural
increases in numbers of women and minority students
obtaining degrees. Possible market adjustments include
the injection of foreign talent, a policy decision to increase
the numbers of women and minority students, and the
recapture and retraining of those engineers no longer in
the field (Collins, 1988).

Figure 1 reveals that only about 5.2 percent of high school sophomores are
likely to pursue studies in the natural sciences and engineering culminating
in receipt of bachelor's degrees in those disciplines. Of those receiving
baccalaureates in 1984, only 4.7 percent will have earned Ph.D.s in science
and engineering by 1992. When these percentages are applied to the cohort of U.S. high school students for the 1986-2000 period, it becomes clear
that the number of young scientists and engineers passing through the education pipeline may not be adequate to meet the demand projected in
Table 1. Planning must be undertaken now to provide the Nation with the
trained personnel who will ensure the development of new technologies and
new knowledge.

Three demographic trends will further complicate the generic issue
of providing a sufficient supply of U.S. scientists and engineers. First, the
18- to 24-year-old cohort that comprises our undergraduate population—
traditionally, whites, both males and females—will continue to decline until

9

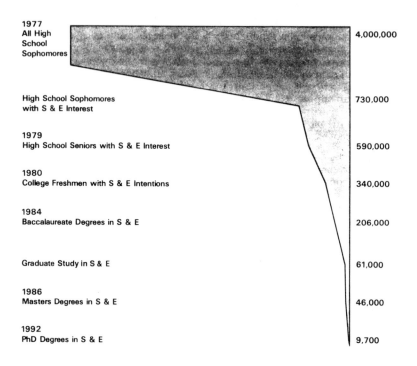

1977 All High School Sophomores	4,000,000
High School Sophomores with S & E Interest	730,000
1979 High School Seniors with S & E Interest	590,000
1980 College Freshmen with S & E Intentions	340,000
1984 Baccalaureate Degrees in S & E	206,000
Graduate Study in S & E	61,000
1986 Masters Degrees in S & E	46,000
1992 PhD Degrees in S & E	9,700

SOURCE: Task Force on Women, Minorities, and the Handicapped in Science and Technology, Changing America: The New Face of Science and Engineering (Interim Report), Washington, D.C.: The Task Force, 1988.

Figure 1. Science and engineering (S&E) pipeline, from high school through Ph.D. degree.

1995 (Figure 2). Second, the percentage of students majoring in most fields of science and engineering has been dropping for the past few years (Table 2). And third, projections show that the increases in the U.S. population will be greatest among ethnic groups that have not heretofore participated significantly in science and engineering. All three trends add to the widespread concern about the future supply of scientists and engineers to meet national needs.

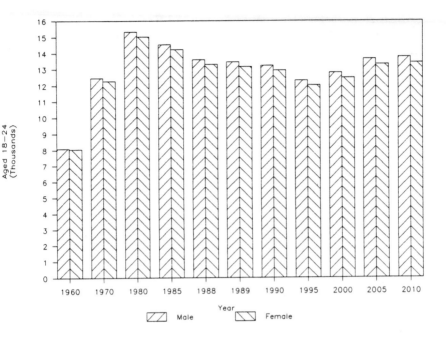

SOURCE: U.S. Bureau of the Census, Statistical Abstract of the United States: 1990 (110th edition), Washington, D.C.: U.S. Government Printing Office, 1990.

Figure 2. U.S. population, aged 18-24, 1970-1988, and projected, 1990-2010 (in thousands).

A predicted consequence is a shortfall of faculty recruits to meet the foreseeable replacement needs due to retirements. Some estimates show that 40 percent of tenured S&E faculty will retire by 1995 and that many new faculty hires will be needed (Figure 3) (Vetter, 1989). Coupled with a decrease in the number of U.S. citizens earning doctorates in science and engineering and gaining tenure—particularly in engineering, computer science, and mathematics—these data project a disturbing picture before the start of the twenty-first century (Thurgood and Weinman, 1990).

11

TABLE 2: Bachelor's Degrees in Science and Engineering as Percentage of All Baccalaureates Awarded, Selected Years, 1972-1989

	Science and Engineering Disciplines							
Year	Comp/ Info	Engng	Health	Life	Math	Phys	Psych	Social
1972	na	4.91	3.00	5.71	2.91	2.23	4.63	9.62
1974	na	4.28	5.29	6.71	2.61	2.09	5.14	9.16
1976	na	3.88	6.42	7.67	2.16	2.14	5.00	8.15
1978	.78	4.77	6.71	7.76	1.36	2.33	4.53	7.59
1980	1.20	5.88	6.84	7.11	1.22	2.35	4.22	7.18
1982	2.13	6.96	6.77	6.64	1.22	2.50	4.26	7.33
1984	3.30	7.86	6.65	5.79	1.36	2.44	4.14	7.07
1986	4.24	7.67	6.66	5.35	1.65	2.17	4.07	6.90
1987	4.00	9.39	6.38	3.84	1.66	2.01	4.32	9.70
1988	3.48	8.94	6.05	3.70	1.60	1.79	4.53	10.09
1989	3.01	8.38	5.81	3.55	1.50	1.69	4.77	10.58

SOURCE: Betty M. Vetter, Professional Women and Minorities (9th ed.), Washington, D.C.: Commission on Professionals in Science and Technology, 1991, from U.S. Department of Education, National Center for Education Statistics, "Degrees and Other Formal Awards Conferred."

Another factor affecting science policy in this area arises from the increasing proportion of both foreign scientists and engineers in the U.S. work force and foreigners earning doctorates from U.S. institutions, which has grown steadily since 1975 (NSF, 1987; Dybas, 1990). The increase in non-U.S. citizens on engineering faculties of U.S. universities (Figure 4) has been especially rapid, but the number of foreign scientists and engineers in the U.S. industrial work force has also been growing steadily as increasing percentages of foreigners receive advanced S&E degrees in the United States (Table 3). Many within the scientific community feel that without

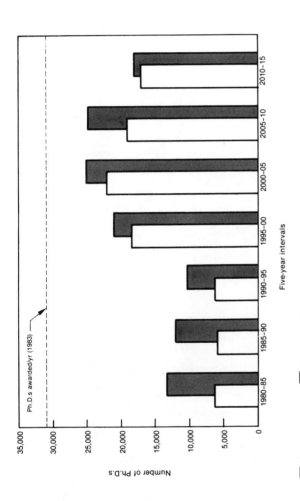

SOURCE: William G. Bowen, *Report of the President* (Princeton, NJ: Princeton University, April 1981); and Office of Technology Assessment analysis.

SOURCE: U.S. Congress, Office of Technology Assessment, *Demographic Trends and the Scientific and Engineering Work Force——A Technical Memorandum*, Washington, D.C.: U.S. Government Printing Office, 1985.

Figure 3. High and low estimates of the number of new Ph.D. faculty hires in the sciences and engineering, every five years, 1980-2015.

13

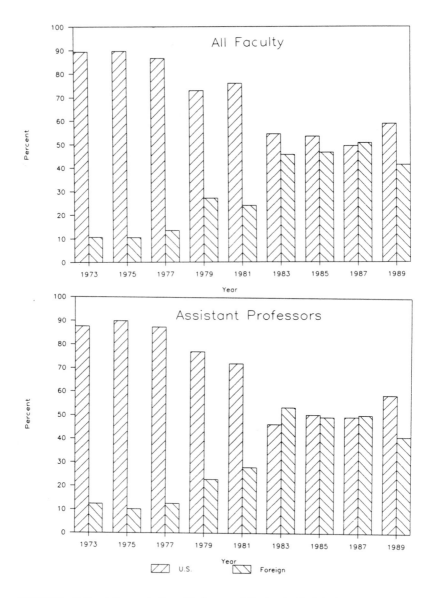

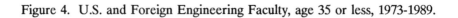

SOURCE: National Research Council's Survey of Doctorate Recipients.

Figure 4. U.S. and Foreign Engineering Faculty, age 35 or less, 1973-1989.

14

TABLE 3: Ph.D.s Awarded by U.S. Universities to Non-U.S. Citizens, 1989

| Field | Total Ph.D.s in Field | Percent Earned by Non-U.S. Citizens | |
		Perm. Visas	Temp.Visas
Total All Fields	34,319	5.1	21.0
Physical Sciences	5,460	5.3	30.5
Physics/Astronomy	1,278	5.3	36.7
Chemistry	1,971	4.6	25.1
Earth, Atmos., and Marine	738	4.4	17.8
Mathematics	861	4.5	44.5
Computer Sciences	612	9.9	31.2
Engineering	4,536	8.7	46.5
Life Sciences*	6,343	4.4	19.3
Biological Sciences	4,106	4.6	15.5
Health Sciences	985	2.8	15.2
Agricultural Sciences	1,252	4.8	35.6
Social Sciences*	5,955	4.2	15.5
Political Sci./Int'l Relations	524	9.3	25.2
Economics	898	6.8	40.8
Humanities	3,558	6.4	10.5
Education	6,265	2.8	7.6
Professional/Other*	2,202	6.3	19.8
Business and Management	1,071	6.9	26.5

Note: Totals in each field include U.S. citizens and recipients with unknown citizenship status. Percentages are based on the number of doctorates with known citizenship status.

* Totals include other fields not shown.

SOURCE: Delores H. Thurgood and Joanne M. Weinman, Summary Report 1989: Doctorate Recipients from United States Universities, Washington, D.C.: National Academy Press, 1990.

the large number of foreign graduate students, U.S. universities would be unable to educate the next generation of scientists and engineers to meet U.S. research and development needs.

The 1990 Immigration Act, signed into law by President Bush in November 1990 to be effective in October 1991, permits 140,000 skilled workers to obtain permanent visas each year. These skilled workers include:

- 40,000 priority workers, those "whose work will prospectively benefit the U.S. and whose achievements are of international renown" as well as "the outstanding researcher or professor . . . recognized internationally and [having] at least three years' teaching or research experience in an academic area;"
- 40,000 immigrants holding advanced degrees;
- 40,000 immigrants with bachelor's degrees or equivalents; and
- 10,000 foreign scientist-entrepreneurs who, in order to qualify, must guarantee a minimum investment of $1 million "in a new commercial enterprise that employs at least 10 U.S. workers" (Eisner, 1991).

The availability of foreign scientists and engineers may be decreasing, however: F. Karl Willenbrock, former assistant director of NSF's Directorate for Scientific, Technological, and International Affairs, noted just a year ago:

> By the end of the century . . . as a group, 29 industrialized and newly industrialized countries that are major producers of [S&E] degrees will experience a decline in college-age people. Many of these are the very countries that in the past have sent science and engineering graduates with baccalaureate degrees to the U.S. for graduate education (quoted in Vetter, 1990).

In addition, Betty M. Vetter, executive director of the Commission on Professionals in Science and Technology, reported that while "13,300 scientists and engineers immigrated to the U.S. in 1970, accounting for 3.6 percent of all immigrants admitted that year, . . . in 1988, their numbers had dropped to 10,900, or 1.7 percent of all immigrants" (Vetter, 1990).

The Ph.D. attainment rate of women remains lower than of men in all fields of science and engineering except psychology, although "at no stage in the educational process is there an indication that the attrition is caused by lack of academic performance" (Koshland, 1988). Recent statistics highlight this attrition by women. In 1986, women received 38 percent of the awarded baccalaureates, 30 percent of the master's degrees, and 26 percent of the doctorates in science and engineering. Little improvement was found three years later, when the percentage of S&E degrees awarded to women were 39, 32, and 26 percent, respectively (Table 4).

These 1989 totals, however, conceal considerable variation by field. Women received 71 percent of bachelor's degrees in psychology, 50 percent in the life sciences (a 7 percent increase from 1986), 46 percent in mathematics, and 44 percent in the social sciences, but only 14 percent in engineering, 30 percent in the physical sciences, and 31 percent in computer and information sciences (a 6 percent decrease from 1986). Whereas women in the life sciences and social sciences are likely to complete advanced degrees, they are much less likely to do so in the physical sciences, computer and information sciences, and mathematics, where women represented 41 percent of the 1978 baccalaureates but only 16.6 percent of the 1986 doctorates. (It should be noted that women's share of math doctorates increased to 19.4 percent in 1989.) In the life sciences, on the other hand, the 30.2 percent of the 1986 doctorates earned by women were based on 36 percent of baccalaureates in 1977; and in engineering, women earned almost 7 percent of 1978 bachelor's degrees and the same proportion of 1986 doctorates. In the biological sciences, where women have traditionally comprised a large number of degree-earners, men have a higher probability

17

TABLE 4: Science and Engineering Degrees Granted to Women, by Degree Level, 1986 and 1989

Science and Engineering Field	Total	Baccalaureates		Master's Degrees		Doctorates	
		No. of Women	% of Total	No. of Women	% of Total	No. of Women	% of Total
Total	1986	121,439	37.7	18,298	29.9	4,906	26.1
	1989	133,395	39.2	21,298	31.6	5,482	26.1
Sciences, total	1986	110,123	45.2	15,970	39.9	4,681	30.4
	1989	121,773	47.7	18,112	42.2	5,082	34.0
Physical	1986	6,698	28.1	1,352	23.3	605	16.4
	1989	5,107	29.7	1,533	26.7	759	19.7
Mathematical	1986	7,036	46.1	1,011	35.0	121	16.6
	1989	7,016	46.0	1,366	39.9	171	19.4
Computer & Info. Science	1986	14,431	36.9	2,037	28.7	49	12.3
	1989	9,416	30.7	2,623	27.9	81	15.1
Life	1986	25,149	43.5	3,491	39.9	1,448	30.2
	1989	18,109	50.2	2,449	49.6	1,298	36.7
Psychology	1986	27,422	68.2	5,417	63.9	1,564	50.9
	1989	34,335	70.8	5,780	67.4	1,834	56.2
Social	1986	29,387	43.5	2,662	37.8	894	32.5
	1989	47,790	44.4	4,361	40.2	939	32.6
Engineering and Engineering technologies	1986	11,316	14.5	2,328	11.0	225	6.7
	1989	11,622	13.6	3,186	13.0	400	8.8

SOURCE: National Science Board, *Science Indicators—1989* (NSB 89-1), Washington, D.C.: U.S. Government Printing Office, 1989; and Delores H. Thurgood and Joanne M. Weinman, *Summary Report 1989: Doctorate Recipients from U.S. Universities*, Washington, D.C.: National Academy Press, 1990.

18

than women of pursuing graduate study, but the difference between men and women is smaller than in many other fields: in 1985, 23.6 percent of the 93,000 recent male undergraduates enrolled in graduate school, as opposed to only 18.1 percent of the women (Hornig, 1987).

Within this environment, current data indicate the need for a concerted effort to analyze the reasons underlying the decreasing participation of U.S. students in science and engineering and to take corrective action. The declining number of college-aged students during the coming decades does not necessarily imply that the United States will have a shortage of native-born scientists and engineers, if a strategy can be found to increase the probability that young people go into scientific and engineering careers. Such a strategy should include increasing the participation of groups who in the past have been underrepresented in the S&E work force. Women are a major human resource that has traditionally been underrepresented in most fields of science and engineering in the United States. Thus, U.S. women educated in the sciences and engineering represent a potential resource for addressing projected future needs for S&E personnel.

Employment Issues

Examination of the 1988 data reveals that, while women are increasingly represented in the total U.S. work force (45 percent) and in the total professional work force, including the scientific, business, and management areas (50 percent), they are greatly underrepresented in the scientific work force (30 percent) and the engineering work force (4 percent) (Figure 5) (NSF, 1990b). In some subfields women are grossly

19

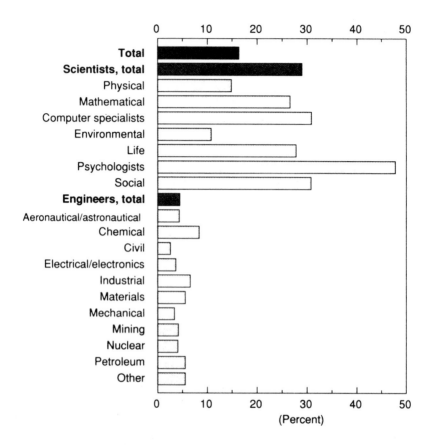

SOURCE: National Science Foundation, Women and Minorities in Science and Engineering (NSF 90-301), Washington, D.C.: U.S. Government Printing Office, 1990.

Figure 5. Percentage of women among employed scientists and engineers, by field, 1988.

underrepresented, more so than in many foreign countries, as was revealed by a recent international study of the participation of women in physics (Table 5). Yet it is projected that by the year 2000, 85 percent of new entrants to the U.S. work force will be women and members of racial/

TABLE 5: Degrees to Women in Physics and Women as Physics Faculty
(in percent)

Country	Degrees to Recent Graduates		Faculty
	Bachelor's	Doctorate	
Belgium	33	29	11
Brazil	24	31	18
Democratic German Republic	12	18	8
France	24	21	23
Hungary	50	27	47
India	25	26	10
Ireland	22	20	7
Italy	29	21	23
Japan	7	4	6
Korea	20	5	3
Netherlands	20	4	6
New Zealand	10	11	6
Philippines	28	60	31
Poland	14	17	17
South Africa	24	21	9
Spain	17	21	16
Turkey	38	17	23
Union of Soviet Socialist Republics	34	25	30
United Kingdom	16	12	4
United States	15	9	3

SOURCE: W. J. Megaw, *Gender Distribution in the World's Physics Departments*, paper prepared for the meeting, Gender and Science and Technology 6, Melbourne, Australia, July 14-18, 1991.

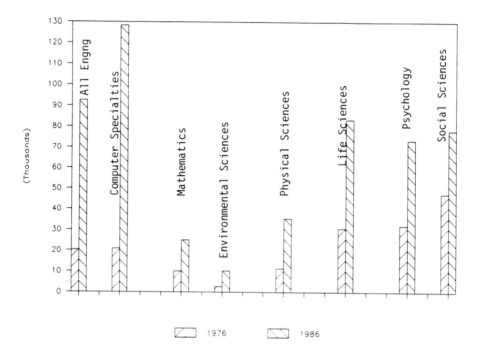

SOURCE; National Science Board, Science Indicators—1989 (NSB 89-1), Washington, D.C.: U.S. Government Printing Office, 1989.

Figure 6. Women doctorates in science and engineering, by field, 1976 and 1986.

ethnic minority groups, groups not traditionally employed in the sciences and engineering (Department of Labor, 1990).

Analysis of their current distribution by fields (Figure 6) and types of employers (Table 6) provides some perspective on the role of doctorate women in the S&E work force:

• **Academe:** Table 6 shows that most women Ph.D.s entered

TABLE 6: Employers of Doctorate Recipients in Science and Engineering, by Sex, 1989

Type of Employer	Year	Total	Male Number	Male Percent	Female Number	Female Percent
TOTAL	1989	476,340	393,843	100.0	82,497	100.0
Self-Employed	1989	31,801	23,216	5.9	8,585	10.4
Business & Industry	1989	111,375	101,097	25.7	10,278	12.5
Academe	1989	225,803	183,901	46.7	41,902	50.8
Two-Year College	1989	5,226	4,006	1.0	1,220	1.5
Medical School	1989	31,711	23,047	5.9	8,664	10.5
Four-Year College	1989	31,693	25,208	6.4	6,485	7.9
Other University	1989	153,154	129,280	32.8	23,874	28.9
Precollege	1989	4,019	2,360	0.6	1,659	2.0
Government*	1989	38,493	32,801	8.3	5,6921	6.9
Nonprofit Org.	1989	13,480	10,429	2.6	3,051	3.7
Other	1989	18,033	12,568	3.2	5,465	6.6
Not Employed	1989	36,495	29,164	7.4	7,331	8.9
No Report	1989	860	667	0.2	193	0.2

*Federal, state, and local.
SOURCE: Office of Scientific and Engineering Personnel, Survey of Doctorate Recipients.

academe (about 50 percent in 1989), primarily medical schools and four-year colleges, with correspondingly smaller numbers finding positions in other types of employment. Another 2 percent of women Ph.D.s entered precollege teaching, compared with less than 1 percent of men. Women comprise 27.6 percent of all faculties at U.S. universities (Vetter, 1991, Tables 5-12), but only 17.5 percent of all science and engineering faculty (Table 7). And while men Ph.D.s are more likely to hold full or associate professorships, women are much more likely to be instructors, lecturers, adjunct faculty, and "other" faculty.

- **Industry:** Overall, about 12 percent of women scientists and engineers are employed in industry, compared to about 26 percent of men scientists and engineers (Table 6). The National Research Council (1983) reported that 1981 data showed a doubling of the number of women scientists and engineers in industry since 1977, but that they remained seriously underrepresented compared to their availability and were underemployed and underpaid. Data from NSF reveal similar findings for the past decade.[1]

- **Government:** In 1988 the U.S. work force included approximately 2.0 million scientists and 2.6 million engineers, of whom 88,106 scientists and 107,415 engineers were employed by the federal government, the largest single employer of scientists and engineers in the United States (Campbell and Dix, 1990). Overall, women and minorities find greater employment opportunities within the

---

[1] The National Science Foundation (1988, p. viii) released the following information: "If those working involuntarily in either part-time or non-S/E jobs are considered as a proportion of total employment, about 6 percent of women, compared with 2 percent of men, are underemployed. . . . Women's salaries are lower than men's in essentially all S/E fields and at all levels of professional experience."

TABLE 7: Academic Ranks of all U.S. Doctorate Recipients in Science and Engineering, 1989

Academic Rank	Year	Total	Male Number	Percent	Female Number	Percent
TOTAL	1989	221,784	181,541	81.9	40,243	18.1
Faculty, Total	1989	199,081	164,254	82.5	34,827	17.5
Professor	1989	89,821	82,354	91.7	7,467	8.3
Assoc. Professor	1989	50,314	40,724	80.9	9,590	19.1
Asst. Professor	1989	38,513	27,235	70.7	11,278	29.3
Instructor	1989	2,445	1,473	60.2	972	39.8
Lecturer	1989	2,395	1,430	59.7	965	40.3
Adjunct Faculty	1989	3,744	2,476	66.1	1,268	33.9
Other Faculty	1989	11,849	8,562	72.3	3,287	27.7
Postdoctoral Appt.	1989	11,892	8,491	71.4	3,401	28.6
Does Not Apply	1989	4,364	3,035	69.5	1,329	30.5
No Report	1989	6,447	5,761	89.4	686	10.6

SOURCE: Office of Scientific and Engineering Personnel, Survey of Doctorate Recipients

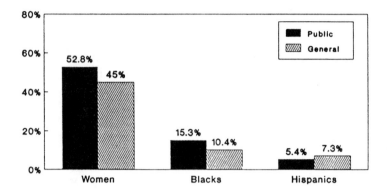

SOURCE: Bureau of Labor Statistics, Employment and Earnings and Labor Force Statistics from the Current Population Survey, in Public Employees: Facts at a Glance, Washington, D.C.: AFL-CIO Public Employee Department, 1990.

Figure 7. Women, blacks, and Hispanics in the federal work force, 1988 (in percent).

federal government than within other U.S. employment sectors (Figure 7). However, "in 1988 only about 14 percent of federal scientists and engineers were female, and about 7 percent were black or Hispanic" (Falk, 1990).

Conclusion

The policy environment for recruiting and retaining women in science and engineering can be characterized currently by attention to three types of issues: (1) demographic considerations, including the rising proportion of non-U.S. citizens in the U.S. work force; (2) education issues, with emphasis on the low rate of participation of women in the component fields of science and engineering; and (3) employment conditions in the U.S. work force. The decisions that we make about our

S&E cadre today will have a significant effect on our ability to find solutions to future problems. Our ultimate success depends upon the degree to which we maximize use of all of the Nation's human resources.

Based on this policy environment, the Committee concludes that there is great potential for increasing the number of women in science and engineering, especially in areas where they are most underrepresented and where the national need is greatest. CWSE has therefore formulated a plan of action on three topics:

1. strengthening the **S&E education infrastructure,**
2. examining the effectiveness of **intervention programs** in sustaining the flow of women into science and engineering, and
3. exploring **career patterns** for women in S&E employment.

Within these topics are four policy issues that the Committee believes warrant consideration:

1. **changing demographics,**
2. **changing missions** of relevant agencies in the federal government and industrial organizations and their subsequent impact on university-government-private partnerships,
3. **changing public attitudes** toward science and scientists, and
4. the **entry and retention of women** into mainstream science and engineering careers.

The next three chapters of this report examine each of the three topics with these policy issues in mind.

2

THE SCIENCE AND ENGINEERING EDUCATION INFRASTRUCTURE

The education system is the most effective way to attract people into a career. As noted in its 1991 Strategic Plan, the Office of Scientific and Engineering Personnel (OSEP) is concerned about the nature of education infrastructure in the United States, which

> has a profound effect on the number and quality of individuals in the science and engineering talent pool. Policies addressing the education infrastructure in the United States are diverse and distributed throughout federal, state and local governments, not to mention the private sector. . . . NRC and OSEP can make a unique contribution to our understanding of the complex issues to be faced by . . . education in the next decade and beyond. While these issues are of interest to many other organizations, few of these other actors effectively link fundamental research with policy formulation. NRC and the broader Academy complex specialize in developing such linkages through its unique committee process.

It is in this linking role that the Committee on Women in Science and Engineering addresses those aspects of the S&E education infrastructure that can increase the participation of women in science and engineering.

Data from the National Center for Education Statistics' National Longitudinal High School Study of the Class of 1972 (NLS-72) and its follow-up studies show that, after expressing an initial interest in S&E studies, individuals often switch to nonscience or nonengineering fields (see, for instance, Burkheimer and Novak, 1981, and Eagle et al., 1988). Many undergraduate S&E majors of both sexes switch to education, law, business, or medicine and other health-related fields for graduate study. For example, of those female freshmen enrolling in engineering programs in 1985, 35.6 percent dropped out of engineering during their sophomore year compared with approximately 16 percent of the male freshman engineering majors (Engineering Manpower Commission, 1987).

The S&E education infrastructure has both formal and informal mechanisms for attracting and retaining talented and qualified individuals into careers in the sciences and engineering. Forming the backbone of the formal S&E education infrastructure are (1) the institutions providing the education to potential scientists and engineers and (2) the policies and programs providing the financial assistance essential for acquiring that education. Informal aspects of the education infrastructure include the media, parents, role models, and mentors. We discuss below the formal and informal mechanisms that have been developed and the data that indicate their effectiveness.

Formal Mechanisms

Various studies have shown that females intending to major in science, mathematics, and engineering have higher attrition rates from those fields than their male counterparts. For instance, a 1986 survey revealed that only 44.4 percent of females (compared with 54.2 percent of males) intending to major in one of those fields actually received a degree in them (Tables 8 and 9). Further,

> an examination of college majors . . . demonstrates that females of all races consistently majored in science, engineering, or mathematics less often than males. . . . White females majored in these fields about half as often

TABLE 8: College Major Field of Study of 1980 High School Seniors Who Had Graduated from College by 1986, by Intended Field of Study in High School and by Sex (in percent)

Sex and Intended Field of Study in College in 1980	College Graduates, by Major Field of Study				All 1980 High School Seniors Planning to Attend College
	Total (sample size)	SEM*	Other	All	
Males					
Total	100.0 (668)	33.9	66.1	100.0	100.0
SEM*	100.0 (230)	54.2	45.8	39.1	31.2
All other fields	100.0 (634)	20.9	79.1	60.9	68.8
Females					
Total	100.0 (786)	18.2	81.8	100.0	100.0
SEM*	100.0 (152)	44.4	55.6	18.8	19.6
All other fields	100.0 (634)	12.1	87.9	81.2	80.4

*Science, engineering, and mathematics

SOURCE: U.S. Department of Education, National Center for Education Statistics, "High School and Beyond" survey, 1986, in Henry A. Gordon, *Who Majors in Science? College Graduates in Science, Engineering, or Mathematics from the High School Class of 1980* (NCES 90-658), Washington, D.C.: U.S. Government Printing Office, 1990.

31

TABLE 9: 1980 High School Seniors Who Graduated from College by 1986, by Major Field of Study and by Race/Ethnicity and Sex

Sex and Race/Ethnicity	College Graduates, by Major Field of Study			Percentage of All 1986 College Graduates	Percentage of All 1980 High School Seniors
	Total (sample size)	SEM*	Other		
Males					
Total	100.0 (730)	30.8	69.2	100.0	100.0
White	100.0 (491)	31.1	68.9	91.2	79.9
Black	100.0 (114)	26.3	73.7	5.1	10.6
Hispanic	100.0 (125)	29.0	71.1	3.8	9.5
Females					
Total	100.0 (868)	16.5	83.6	100.0	100.0
White	100.0 (575)	15.7	84.3	88.5	78.8
Black	100.0 (161)	23.8	76.2	7.8	12.2
Hispanic	100.0 (132)	18.1	81.9	3.7	9.0

*Science, engineering, or mathematics.

SOURCE: U.S. Department of Education, National Center for Education Statistics, "High School and Beyond" survey, 1986, in Henry A. Gordon, *Who Majors in Science? College Graduates in Science, Engineering, or Mathematics from the High School Class of 1980* (NCES 90-658), Washington, D.C.: U.S. Government Printing Office, 1990.

as white males (15.7 percent versus 31.1 percent). Black
females, however, majored in science, engineering, or
mathematics almost as often as black males (26.3 percent
of males versus 23.8 percent of females) (Gordon, 1990;
Table 9).

The near parity of black females with black males suggests that further
study should be done to probe cultural and sociological reasons for this
result.

Institutions

Analyzing the status of women in the S&E education pipeline, one
must examine those institutions most effective in producing women scientists and engineers and the programs they have in place to achieve that
goal:

- **Ph.D.s**: As shown in Table 10, the 10 U.S. doctorate-granting
  institutions that awarded the most S&E degrees during the past
  decade are University of California-Berkeley, University of Illinois-
  Urbana/Champaign, Massachusetts Institute of Technology (MIT),
  University of Wisconsin-Madison, Cornell University, Stanford
  University, University of Minnesota-Minneapolis, Purdue
  University, University of Michigan, and University of California-
  Los Angeles. When ranked by S&E Ph.D.s awarded to women,
  however, their rank order changes dramatically (see Related Tables
  A and B), and MIT and Purdue are displaced in the top 10 by the
  Ohio State University and the University of Maryland. These

TABLE 10: Top 25 Science and Engineering Doctorate-Granting Institutions, 1980-1990 (all graduates)

Institution	1980	1981	1982	1983	1984	1985	1986	1987	1988	1989	1990	Total 1980-1990
TOTAL, MALE AND FEMALE												
Calif, U-Berkeley	540	483	542	519	517	549	563	534	585	658	607	6097
Ill, U, Urbana-Champ	394	411	358	382	381	442	383	434	444	468	514	4611
Mass Inst Technology	372	384	391	412	389	421	442	436	489	469	478	4683
Wisconsin, U-Madison	400	389	442	398	399	458	413	453	479	485	460	4776
Cornell Univ/NY	330	333	328	355	351	337	371	367	377	395	455	3999
Stanford Univ/CA	327	358	339	308	364	334	393	405	411	412	411	4062
Minnesota, U-Minneapl	291	316	291	275	312	336	381	310	337	360	403	3612
Purdue University/IN	297	326	288	305	308	308	320	300	302	348	383	3485
Michigan, Univ of	292	315	323	376	348	371	348	344	344	335	382	3778
Calif, U-Los Angeles	314	330	310	311	296	296	282	288	363	334	382	3506
Texas, U-Austin	219	228	234	228	227	255	298	330	326	329	367	3041
Ohio State Univ	300	274	305	293	269	323	297	322	307	371	366	3427
Texas A&M University	211	195	180	202	227	221	228	257	253	310	305	2589
Maryland, Univ of	175	172	202	192	209	210	213	220	205	244	301	2343
Michigan State Univ	281	279	305	299	250	240	242	258	268	287	286	2995
Washington, U of	228	231	246	249	237	221	249	262	279	272	283	2757
Penn State Univ	215	233	240	261	243	234	237	251	260	289	282	2745
Florida, Univ of	169	166	142	206	203	216	203	223	236	259	273	2296
Harvard Univ/MA	248	232	244	256	246	205	243	215	242	221	269	2621
NC State U-Raleigh	112	128	168	164	171	175	193	180	210	199	252	1952
Columbia University	228	231	222	197	233	245	219	215	223	254	243	2510
Pennsylvania, U of	201	214	256	217	211	206	202	248	221	272	240	2488
Northwestern Univ/IL	178	184	193	179	188	218	207	211	220	259	234	2271
Calif, U-Davis	233	253	193	276	239	209	229	234	252	247	234	2599
Arizona, Univ of	184	154	179	188	197	182	171	214	225	237	228	2159

SOURCE: National Science Foundation, unpublished data.

results may indicate that other universities are taking steps to broaden the supply of female Ph.D.s in S&E fields.

- **Baccalaureate Origins:** Many of the same institutions that are successful in retaining female S&E graduate students to completion of doctorates have also provided their undergraduate education in S&E fields. Data from the Doctorate Records File indicate that efforts in this area during the 1985-1990 period were particularly successful at University of California-Berkeley, Cornell University, University of Michigan, University of California-Los Angeles, University of Illinois-Urbana/Champaign, and University of Wisconsin-Madison. Joining those institutions to form the top 10 baccalaureate institutions of women who received S&E Ph.D.s during 1985-1990 were Pennsylvania State University, Rutgers University, University of California-Davis, and the University of Pennsylvania. The latter four institutions also awarded 17 percent of the Ph.D.s granted to women in the sciences and engineering by the top 25 institutions between 1985 and 1990. As shown in Table 11, however, the number of doctorates awarded to women who received undergraduate degrees from the same institution varies by field. Data from NCES (1970 +) confirm that women, particularly minority women, are somewhat less likely than men to attend the most prestigious research universities as either undergraduate or graduate students.

Availability of Financial Support

Financial aid is a very important factor in recruiting and retaining able women in science and engineering. At the undergraduate level, schol-

TABLE 11: Top Five Baccalaureate Institutions of Female Science and Engineering Doctorate Recipients, by Field of Doctorate, 1985-1990

Physical Science	Agriculture	Biological Science	Psychology	Social Science	Engineering
1 UC-Berkeley	1 Cornell Univ.	1 Cornell Univ.	1 UC-Los Angeles	1 UC-Berkeley	1 Univ. of Illinois, Urbana-Champlain
2 Cornell Univ.	2 Univ. of Illinois, Urbana-Champlain	2 UC-Berkeley	2 Univ. of Michigan	2 Univ. of Michigan	2 UC-Berkeley and Purdue Univ.
3 Wellesley College	3 UC-Davis	3 UC-Davis	3 UC-Berkeley	3 UC-Los Angeles	3 Univ. of Michigan and Penn State
4 Univ. of Michigan	4 Michigan State	4 Univ. of Michigan	4 Cornell Univ.	4 Univ. of Wisconsin, Madison	4 Cornell Univ.
5 Rutgers Univ.	5 Univ. of Wisconsin, Madison	5 Univ. of Illinois, Urbana-Champlain	5 Univ. of Wisconsin, Madison	5 Univ. of Minnesota, Minneapolis	5 Ohio State

SOURCE: National Research Council, Doctoral Records File, unpublished data.

arships to women for S&E studies often reinforce recruitment efforts (NSF, 1990b; Moran, 1986). Furthermore, undergraduate women are encouraged to continue their S&E studies because they know that financial support will be available for continued studies at the graduate level. At the graduate level, recruitment is strongly tied to the availability of financial support (see, for instance, Anderson, 1990), and retention requires consistent, continuing support.

However, women do not receive the same kinds and levels of financial aid as their male counterparts in science and engineering (Table 12), and this may inhibit their entry. An increase in the probability that women students will receive financial support could yield significant increases in female participation in the undergraduate and graduate student segments of the pipeline (Coyle, 1986). Research indicates that women who are offered financial aid at the beginning of their undergraduate education are more likely to continue their studies in the sciences and engineering (Rosenfeld and Hearn, 1982). In addition, needy students, those who cannot afford to complete their education without interruption to earn more money, may require special alternatives such as part-time or continuing education programs, perhaps developed in cooperation with industry. The availability of sustained financial aid when needed by students later in their undergraduate education is also important for retention (Connelly and Porter, 1978).

Variations in Ph.D. attainment rates by S&E field are highly correlated with the availability of financial support (Tuckman et al., 1990). Some universities have responded favorably to this finding: Yale University, for instance, has decreased the use of teaching assistants (TAs) but now encourages graduate students to earn Ph.D.s more rapidly by

TABLE 12: Percentage Distribution of Primary Sources of Support of Doctorate Recipients, by Sex and Broad Field, 1989

Source/Gender	Year	Total Fields	Phys. Scncs.	Engrng. Scncs.	Life Scncs.	Social Scncs.	Human.	Prof/ Educ.	Other
Personal									
Men	1989	34.1	13.7	15.3	22.3	49.1	48.0	74.6	53.5
Women	1989	51.1	13.0	12.5	27.3	59.5	48.0	77.6	57.2
Federal, Non-R.A.									
Men	1989	5.3	4.1	4.1	13.0	4.0	2.3	2.7	2.1
Women	1989	5.7	4.3	10.4	15.1	5.2	1.5	1.9	1.3
R.A., Fed. & Univ.									
Men	1989	27.2	45.4	49.7	34.4	9.2	1.5	3.0	7.2
Women	1989	15.1	42.8	50.5	30.8	8.5	1.5	3.8	9.2
Teaching Assistant									
Men	1989	17.5	25.9	12.1	10.9	21.7	31.5	5.9	21.2
Women	1989	15.7	29.5	10.0	11.8	14.6	35.4	6.4	19.1
Fellowship									
Men	1989	6.0	4.7	4.7	7.7	7.6	11.3	2.4	4.9
Women	1989	6.0	4.9	9.7	8.1	7.1	9.1	2.4	5.2
Other Sources									
Men	1989	9.9	6.2	14.1	11.6	8.4	5.4	11.3	11.1
Women	1989	6.4	5.4	6.9	6.8	5.2	4.4	7.7	8.0

SOURCE: Delores H. Thurgood and Joanne M. Weinman, *Summary Report 1989 Doctorate Recipients from U.S. Universities*, Washington, D.C.: National Academy Press, 1990.

offering them fellowships to finish their dissertations (Cheney, 1990).
However, it was pointed out to the committee that

> The graduate education process is evolving into a system
> serving the needs of the faculty and institution at the
> expense of the needs of the graduate student population.
> The shrinking availability of research funds accelerates this
> process, further compromising the quality of the graduate
> experience. All graduate students are adversely affected,
> but women in graduate programs are especially impacted
> because of their traditional lack of assertiveness. . . . Their
> dependency on a major advisor for financial support may
> force them to endure misuse or abuse: long hours in the
> laboratory, excessive teaching responsibilities, extended
> stays in the graduate program (Mulnix, 1990).

Table 12 shows that women graduate students in the life sciences and the
social sciences are more likely than men to be self-supporting and less
likely, in general, to be funded as either TAs or research assistants (RAs).
Thus, relative to men, women overall are more likely to be deprived of
research time and important opportunities for interaction with peers and
faculty.

The extent to which these problems occur varies by field and by
race/ethnicity. In this context, OSEP examined the numbers of women
applying for and receiving graduate fellowships in the programs it
administers for NSF. These fellowships are highly selective and prestigious
and are generally regarded as early indicators of future success. Although
women in general have received about one-third of those awards, primarily
in the earth, biomedical, biological, and behavioral sciences—the same
fields in which most women apply (Tables 13 and 14; see also Related
Tables C, D, E, and F)—the percentage of awards to women has increased
steadily since 1985. Overall, the proportion of women receiving NSF
graduate fellowships is lower than among men, though it appears to vary

TABLE 13: NSF Graduate Fellowship Program Applications and Awards, by Sex, 1985 and 1991

Discipline	1985		1991		1985		1991	
	M	W	M	W	M	W	M	W
	Total Applicants				Total Awards			
N	2776	1614	4145	3201	362	178	556	394
%	63.2	36.8	56.4	43.6	67.0	33.0	58.5	41.5
Biochem*	246	167	276	256	32	16	31	31
	59.6	40.4	51.9	48.1	66.7	33.3	50.0	50.0
Biology	298	274	369	432	32	40	40	53
	52.1	42.9	46.1	53.9	44.4	55.6	43.0	57.0
Chemistry	219	118	272	174	32	9	41	16
	65.0	35.0	61.0	39.0	78.0	22.0	71.9	28.1
Earth Sci	151	88	116	124	20	9	13	16
	63.2	36.8	48.3	51.7	69.0	31.0	44.8	55.2
Appl Math/	80	39	100	87	14	1	18	4
Statistics	67.2	32.8	53.5	46.5	93.3	6.7	81.8	18.2
Mathematics	105	43	132	90	19	1	22	10
	70.9	29.1	59.5	40.5	95.0	5.0	68.8	31.2
Physics and	309	44	404	99	39	6	57	13
Astronomy	87.5	12.5	80.3	19.7	86.7	13.3	81.4	18.6
Behavioral	397	436	627	780	50	50	92	89
Sciences**	47.7	52.3	44.6	55.4	50.0	50.0	50.8	49.2
Biomedical	154	208	195	311	15	28	23	30
Sciences	42.5	57.5	38.5	61.5	42.5	57.5	43.4	56.6
Computer	182	54	282	67	27	3	40	5
Science	77.1	22.9	80.8	19.2	90.0	10.0	88.9	11.1
Engineering	635	143	1280	692	82	15	179	127
	81.6	18.4	64.9	35.1	84.5	15.5	58.5	41.5

* Includes biochemistry, biophysics, and molecular biology.
**Prior to 1991, this field included psychology, economics, and sociology. Because the disaggregation of behavioral sciences———into (1) anthropology, sociology, and linguistics; (2) economics, urban planning, and history of science; (3) political science, international relations, and geography; and (4) psychology———did not occur until 1991, a single category is used here.
SOURCE: Office of Scientific and Engineering Personnel.

TABLE 14: NSF Minority Graduate Fellowship Program Applications and Awards, by Sex, 1985 and 1991

Discipline	1985		1991	
	Men	Women	Men	Women
Total Applicants				
N	298	305	595	644
%	49.4	50.6	48.0	52.0
Biosciences*	62	79	93	158
	44.0	56.0	37.1	62.9
Chemistry/	27	22	48	41
Earth Science	55.1	44.9	53.9	46.1
Phys/Astron/	37	32	82	56
Math	53.6	46.4	59.4	40.6
Behavioral	68	116	113	207
Science**	37.0	63.0	35.3	64.7
Engineering	65	35	172	119
	65.0	35.0	59.1	40.9
Total Awards				
N	39	21	87	63
%	65.0	35.0	58.0	42.0
Biosciences*	10	5	16	13
	66.7	33.3	55.2	44.8
Chemistry/	2	2	4	4
Earth Science	50.0	50.0	50.0	
Phys/Astron/	6	1	9	8
Math	85.7	14.3	52.9	47.1
Behavioral	12	11	18	21
Science**	52.2	47.8	46.2	53.8
Engineering	9	2	40	17
	81.8	18.2	70.2	29.8

* Includes biology, biochemistry, biophysics, and biomedical science.
**Includes anthropology, sociology, and linguistics; economics, urban planning, and history of science; political science, international relations, and geography; and psychology.
SOURCE: Office of Scientific and Engineering Personnel.

unpredictably from one field to another and from year to year. Women applicants fare particularly poorly in the fields of computer science, applied mathematics/statistics, and physics/astronomy. Until 1991, when they received 31 percent of the awards in mathematics, women received less than 18 percent of the graduate fellowships in that field.

Informal Mechanisms

Informal efforts to recruit women into S&E fields typically:

- address the negative public image of scientists and engineers and of science and engineering;
- encourage precollege interest of young women in S&E majors and careers;
- involve parents and peers; and
- as in formal programs, provide opportunities for female students to interact with scientists and engineers in academe, industry, and government who serve as role models and mentors.

Research on retention of both men and women in undergraduate S&E programs indicates that effective programs include the following: orientation programs for freshmen, remedial courses, career seminars, educational and career counseling, peer tutoring, research opportunities, cooperative and summer job programs, campus chapters of professional organizations such as the Society of Women Engineers, recognition awards and events, and exit interviews with graduating seniors. Successful retention programs, such as Purdue University's Women in Engineering Program and the Women in Science Program of Rutgers University's

Douglass College, have used two other intervention actions that can affect retention of women in S&E majors:

1. the use of professional counselors with training both in the special problems faced by undergraduate women in traditionally "masculine" fields of study and in specific counseling strategies that can increase women's persistence in these fields; and

2. interactions with industrial scientists and engineers in order to enhance the motivation of beginning S&E students (LeBold, 1987).

Two additional factors affecting undergraduate retention were noted in the National Engineering Career Development Study: academic performance during the freshman year; and self-perceptions of math, science, and problem-solving ability (Shell et al., 1985). These same factors could also be applicable to undergraduate science majors.

The Role of the Media

In order to recruit male or female students into science and engineering, those fields must be perceived as positive career choices (MacCorquodale, 1984). However, a number of recent studies in various developed countries suggest that science and engineering, in general, have an "image problem." When students and adults are asked about their image of scientists and engineers, not only are science and engineering strongly viewed as traditionally masculine fields of study, but in most cases scientists and engineers are pictured as "mad" scientists and perpetrators of destruction (Kahle and Matyas, 1987).

The positive benefits of S&E research and development have not been the primary focus of the public image, nor have science and engineering generally been viewed by the public as ennobling careers (OTA, 1988; NAS, 1989). Even a cursory glance at popular television and print materials (such as comic books) suggests that the popular media do little to change this public image and can have an important negative influence on students' images of science and engineering and of scientists and engineers. The potential for using popular media in recruitment strategies remains largely untapped (Task Force, 1988).

Parental Guidance

A study by the American Association for the Advancement of Science found that most of the most effective precollege programs to increase females' participation in science and mathematics involve parents in some way (Malcom, 1983). Parents play an important role in influencing the initial career choices of all students, but especially those of young women. However, there has been no systematic evaluation of programs and materials informing parents about the importance of science and mathematics education for their children, girls as well as boys, and guiding parents on how to assist their children in career choices in these areas.

Role Models and Mentors

Research indicates that students, both male and female, are influenced by role models and faculty members (see, for instance, Nagy and Cunningham, 1990). Opportunities to interact with S&E personnel have long been central to "career day" and other precollege programs designed to spark young women's interests in S&E careers.[2] As John F. Welch, Jr. (1991), chairman and chief executive officer of the General Electric Company, wrote recently:

> Corporate volunteers can guide America's students toward a world of work, study, and achievement. . . . GE volunteers and their counterparts at a few other companies have proved that social and economic upward mobility——the glue that holds us all together——can be restored. American nightmares [about inability to compete in the global marketplace] can be changed into American dreams.

Undergraduate women in science and engineering have been effectively used to recruit high school students, and women graduate students have successfully served as recruiters of women undergraduates in science and engineering (Hall and Sandler, 1983). At present, however, female S&E faculty role models are most likely to be found among the untenured junior faculty and, therefore, are not generally available for significant time commitments to recruiting and other activities involving greater interactions with students (Cheney, 1990). Recruitment of women students at a given institution would be enhanced by the presence of women faculty at all ranks, a signal to women students that they will be respected and treated fairly. The presence of women faculty at junior ranks only or in adjunct or off-ladder status signals the opposite (Sandler, 1986). However, many top graduate departments in science and engineering still

---

[2] For descriptions of some of these programs, see Sandra L. Keith and Philip Keith, eds., *Proceedings of the National Conference on Women in Mathematics and the Sciences* (St. Cloud, Minn.: St. Cloud State University, 1990).

have no tenured women faculty, which gives an even more negative signal (see, for instance, Selvin, 1991).

Undergraduate women and men at large research universities are negatively affected by the frequent lack of interactions with the research-oriented faculty in their departments (Smith, 1990). Many highly talented students may not be receiving adequate encouragement to pursue graduate study. Such a phenomenon would affect women more than men, because women are usually less plugged into the network (Mulnix, 1990). In response, some institutions encourage women S&E faculty members to act as role models and mentors for undergraduate and graduate women in their departments (Malcom, 1983). Institutions address this issue through formal programs that (1) sensitize faculty to the needs of women students, (2) follow the progress of women students throughout their enrollment period, and (3) promote mentoring between undergraduate, graduate, and postdoctoral women in science and engineering. Examples of programs that seem effective are the Illinois Institute of Technology's Women's Mentoring Organization and the University of Chicago's Mellon Instructorships, which "offer new Ph.D.s the opportunity to work with mentors teaching in the common core (Cheney, 1990), as well as the University of Washington's Women in Engineering Initiative.

Institutional Factors

Attrition from S&E majors is seldom related only to academic talent and achievement, especially for women (Roby, 1973; LeBold, 1987; Hall, 1982; Sandler, 1986). As Cavanaugh (1990) noted,

Women often "drop out" of science in graduate school or

46

even after starting their careers. The major factor is the climate of the workplace, with its competitiveness, subtle forms of sexual harassment, off-track assignments or limited responsibilities, and lack of encouragement. Add to this lower salaries and promotion rates, inappropriate responses to reproductive hazards, and lack of provision for child-care and the difficulties of staying in science become obvious.

In addition to formal barriers and overt discrimination, women completing studies in traditionally masculine fields often encounter subtle forms of discrimination called "micro-inequities" (Hall, 1982; Ehrhard and Sandler, 1987) that contribute to an unsupportive "campus climate." On an incident-by-incident basis, micro-inequities appear to be insignificant, but collectively they make an important and significant difference in the collegiate experience of men and women. For example, women who try to participate in classroom discussion are ignored or interrupted more frequently than men by both faculty and male students; their questions are more often treated as trivial by faculty; and they are frequent targets of "good-natured" derogatory humor (Sandler, 1986; Mulnix, 1990). Anecdotal evidence also indicates that faculty, teaching assistants, and graduate students from certain cultures are less accustomed to the presence of female students in the classroom and laboratory and may discriminate against women students either consciously or unconsciously.[3] However, the

---

[3] This issue was a topic of much discussion at the conference, "Women in Science and Engineering: Changing Vision to Reality," of the American Association for the Advancement of Science, July 29-31, 1987, and at meetings of the National Research Council's Committee on the International Exchange and Movement of Engineers [see National Research Council, *Foreign and Foreign-Born Engineers in the United States: Infusing Talent, Raising Issues,* Washington, D.C.: National Academy Press, 1988, and *Engineering Education and Practice in the United States,* Washington, D.C.: National Academy Press, 1985]. Although a recurrent theme during subse-

Committee knows of no research undertaken to determine how common this phenomenon is or how to combat it.

Many academic institutions are unaware of the successful activities by other institutions to create a supportive campus climate. Besides programs mentioned earlier, these include data collection and analysis from each department on the participation and advancement of women at the undergraduate, graduate, and faculty levels. The campus climate for women is also enhanced by on-campus branches of professional societies——such as the Society of Physics Students, Chicanos in the Health Sciences, and the Society of Women Engineers——that promote interactions between S&E professionals and students and shepherd women students into professional careers.

Priority Issues

Policies affecting the S&E education infrastructure are diverse, and many groups——public and private alike——have placed high priority on developing programs to increase the number and quality of women entering science and engineering careers. After some discussion, we have concluded that an effective role for the Committee on Women in Science and Engineering in this area will be:

- stimulating data collection, to assess the effectiveness of

---

quent meetings of various professional scientific organizations, this issue has not yet been studied in depth.

educational programs that have been introduced formally over the years;

* examining data on science majors graduating from historically black undergraduate colleges and universities, to determine the effectiveness of HBCUs in preparing those graduates for S&E careers;

* specifying those features of effective programs developed in one institution that can be duplicated in another;

* collecting data in order to analyze and evaluate the effectiveness of college admissions policies in newly coeducational institutions, some of which are major sources of future S&E Ph.D.s and which may routinely establish quotas for admitting women and racial/ethnic minorities;

* studying the career differences of men and women S&E doctorates, by discipline, with reference to their education;

* developing techniques to disseminate information to academic administrators on the importance of role models and mentors in the undergraduate and graduate S&E infrastructure, pointing out institutional mechanisms that are effective in producing S&E doctorates;

* examining the incentives (financial support, etc.) available for potential S&E majors;

* conducting regional and/or national conferences on the effective partnerships in science and engineering between academe, industry, and government; and

* planning strategic "awareness" sessions for decision makers in the print and visual media in order to eradicate the negative image of science and engineering in society.

3

EFFECTIVENESS OF INTERVENTION MODELS

In spite of growing awareness of the Nation's increasing need for scientists and engineers, few policies or programs have been implemented to attract and retain women in these occupations in substantial numbers. However, a few effective intervention programs have been implemented sporadically throughout the pipeline, beginning at the precollege level and continuing through employment. For purposes of discussion, intervention programs are defined as

> efforts to open up the pathway to science and engineering careers for underrepresented groups. . . . usually a series of activities . . . to address one or more specific factors which affect females' or minorities' interest, motivation, or skills in science, mathematics, or engineering (Matyas and Malcom, 1991).

A recent study of the American Association for the Advancement of Science (AAAS; Matyas and Malcom, 1991) found that, for the six groups tar- geted by intervention programs (precollege, general undergraduate, S&E undergraduate, general graduate, S&E graduate, and faculty), 51 percent of such programs targeted "'minorities only' and 'all' students with special efforts for minorities." By contrast, only 9 percent of the intervention programs targeted "'women only' and 'all' students with special efforts for women," and another 28 percent targeted "'minorities and women' and 'all' students with special efforts for minorities and women." The study found that programs targeting minority students do serve women: 51 percent of those participating in such programs are women, and black women participate to a much higher extent than do Hispanic and American Indian women.

The Committee believes that it can play an effective role in recommending areas where the public and private sectors alike can mount intervention programs and expand successful efforts to identify, educate, and employ talented women in science and engineering. The programs included in this section are examples of work under way—and by no means is this chapter intended to be an exhaustive treatment of intervention models.

Precollege Programs

Most of the programs currently under way to increase the participation of women in science and engineering are directed at the precollege level. For instance, the National Research Council recently sponsored studies of both precollege mathematics and biology (MSEB/ BMS, 1989; NRC, 1989). Professional scientific societies have sponsored a variety of programs for high school students—for example, the American Institute of Physics' Science Education for Equity Reform (SEER) and the American Chemical Society's course, "Chemistry in the Community"—as have state boards of education (Silverstein, 1990; Lee, 1991), private corporations, and universities (Matyas, 1987). In major U.S. cities, it is not uncommon to find a dozen or more intervention programs in which precollege students are actively engaged in science and math activities (Kay, 1990).

The range of precollege intervention programs is broad, including both long- and short-term projects. Activities associated with short-term projects—those usually lasting one or two days—are career information, course information, and role models. In addition to these activities, long-

term projects——those of several consecutive days, weekends, or weeks——
often have the following components: hands-on experience, problem-
solving and test-taking exercises, and discussions of financial aid.

Some federal agencies, such as the U.S. Department of Energy,
have funded summer laboratory experiences for high school students. A
new federal initiative has as one of its goals the expansion of opportunities
in science and engineering for these students: the Committee on Education
and Human Resources of the Federal Coordinating Council for Science,
Engineering, and Technology (FCCSET) has "established strategic objectives and priorities for funding Federal programs in mathematics, science,
engineering and technology education" (FCCSET, 1991) to meet the national education goal that "by the year 2000, U.S. students will be first in
the world in science and mathematics achievement." To that end, the President's fiscal year 1992 budget requested $1.94 billion for mathematics and
science education programs; $660 million was requested for precollege pro-
grams, including an increase of 28.4 percent in funding for precollege programs than enacted for fiscal year 1991. This increase represents two-
thirds of the total increase of $225 million for math and science education.

Undergraduate Programs

Prior to the 1991 AAAS study, there were no comprehensive
studies of interventions targeting women majoring in a science or
engineering discipline at the undergraduate level. However, research on a
limited number of such programs has shown that those most effective——that
is, those most successful in awarding bachelor's degrees in the sciences and
engineering to women——share common elements:

53

- having a developed program of recruitment and retention activities;
- using a variety of admission criteria to predict the performance of female students in S&E study;
- providing opportunities for female students to interact with professionals in academe, industry, and government;
- offering informed undergraduate and graduate counseling about course choices and future job opportunities;
- developing support or networking groups—both informal gatherings and the more structured meetings of student chapters of professional organizations, such as the Society of Women Engineers—to reduce feelings of isolation and alienation for women in traditionally masculine fields; and
- cultivating departmental and campus climates that encourage academic achievement among women.

Undergraduate intervention programs fall into four basic types:

1. general retention programs providing academic support, such as tutoring, study skills courses/centers, and bridge programs "to assist students in the transition from high school to college;"
2. science and engineering recruitment and retention programs, such as scholarships and campus chapters of professional societies;
3. offices of minority affairs and cultural centers whose activities relate to minorities in general or to specific racial/ethnic groups; and
4. offices of women's affairs or women's studies programs (Matyas and Malcom, 1991).

Some of these strategies are currently being tested at institutions across the country—for instance, the Women in Math and Science Program at Rutgers University's Douglass College and at the University of Colorado-Boulder and Purdue University (Keith and Keith, 1990). The Douglass program was begun in 1986 to retain high school, undergraduate, and graduate women who have expressed interest and ability in mathematics and science.[4] The Committee on Women in Science and Engineering learned recently that other institutions of higher education are incorporating aspects of the Douglass program on their campuses. For instance, a 1991 announcement to incoming students at Washington State University noted that

> Stephenson North Residence Hall has now designated a floor for women in math, science, and engineering and a floor for men in math, science, and engineering . . . . Many programs [to] be offered to . . . students include tutorials, study room, study groups, [and] peer advising.

Academic counseling and tutoring, faculty interaction, and research experience are important for both male and female students in science and

---

[4] The Douglass program has 10 components: Bunting-Cobb Residence Hall for 99 undergraduate women majoring in math, science, and engineering and 10 graduate women, who serve as Bunting-Cobb Fellows/mentors to the undergraduates; peer study groups, each led by a graduate student, "whose purpose is to get students to help each other;" peer tutoring by upperclass undergraduates; study partnerships; Douglass Science Management Intern Program, whereby seniors majoring in math, science, or psychology work in scientific or technological corporations for one semester; career panels, held each semester, where women employed as mathematicians, scientists, and engineers can share career information and serve as role models for students; visits to the laboratories of female scientists and engineers, who discuss their research; informal talks with women faculty; Big Sister/Little Sister Program, in which a first-year student may choose to have an upperclass woman as her "big sister;" and distribution of information on research/internship opportunities (Kennedy, 1990; see also Mappen, 1990).

engineering. Ideally, undergraduate counselors are informed about the special problems faced by many women in traditionally "masculine" fields of study, such as lack of appropriate communication skills (i.e., "assertiveness, debating, standing firm, and even bluffing"), and utilize specific counseling strategies that can increase women's persistence in these fields (Mulnix, 1990). However, training programs for undergraduate academic counselors do not always address these issues in detail.

Various studies show that many talented young women experience a loss of self-confidence early in their undergraduate years (Arnold, 1989). This may be mitigated through a variety of mechanisms, including faculty interactions, research experiences, work-study programs, and other opportunities to gain hands-on experience in science or engineering (Brown, 1990; McLaren, 1990).[5] One particular "success story" is Carleton College, Minnesota, where the percentage of women receiving bachelor's degrees in chemistry increased from 18 percent in 1970 to 50 percent in 1991, although the ratio of men to women remained stable:

> [No] rigorous studies have been conducted to determine what produced these changes, . . . [but] the faculty are very important in creating an environment that is supportive of women students. For the past 15 years, at least one of the six faculty members in the chemistry department has been a woman. . . . Students, both male and female, have commented that their presence has sent a message that women can expect to find satisfaction and success in a career in science. . . . There is an expectation that all students, regardless of sex, will be challenged and will rise to meet the challenges presented to them. Gender is

---

[5] An example of a work-study program that makes a special effort to involve women students is the Department of Energy's National Laboratory Cooperative Summer Program.

never used as a factor in the selection of student teaching assistants or research associates, and students in the department are well aware of this policy (Finholt, 1990).

Several private corporations fund programs targeted to under-participating groups. For instance, A&T offers undergraduate scholarship programs in ceramics engineering, chemistry, computer science, mechanical engineering, chemical engineering, physics, mathematics, statistics, and materials science. But interventions directed to potential female scientists and engineers, like many other innovations in science, mathematics, and engineering education, lack visibility and thus are not used by a broader base of educators. For this reason, it is vitally important to disseminate information about effective programs.

Graduate Programs

In general, graduate intervention programs address admissions, continuing education, financial aid, academic programs, student development, counseling, and support services (Bogart, 1984). The recent AAAS study found that graduate programs targeting students in the sciences and engineering fell within five categories:

1. fellowships for research,
2. Health Careers Opportunities Program,
3. graduate recruitment and retention programs in science and/or engineering,
4. seminars for graduate students who will work as teaching assistants, and

5.   bridge programs which assist students in the transition from
        undergraduate to graduate studies (Matyas and Malcom, 1991).

        Recent recommendations for expanding opportunities for women
in the sciences and engineering at the graduate level include federal funding
support for graduate S&E education for women, provision of federal
agency funds as an incentive to researchers to employ female graduate
students, and federal monitoring of progress toward these goals (Task
Force, 1988; OTA, 1988). At the same time, however, the science
community cannot reach consensus about what should have priority in the
federal budget: projects such as mapping the human genome and the
Superconducting Super Collider compete directly with basic research (and
the stipends paid to research assistants), and it is uncertain how Congress
will distribute funds from year to year (OTA, 1990):

> [Big] projects have glamor on their side . . . [but] if they
> get a priority over the less visible work going on in
> university and government labs, one loser will be the
> diversity of the country's scientific competence. Another
> will be the training of young scientists (The Washington
> Post, 1991).

        Research-oriented liberal arts colleges are more effective than the
highly selective research universities at encouraging both female and male
students to pursue scientific and engineering careers (Oberlin College,
1986a & b). This points to a need in the major research universities for
closer faculty contacts, which would benefit not only women but all
students. CWSE learned that some research universities have reacted
positively to this finding. The AAAS study found that "programs designed
to recruit and/or retain women and minorities in science and engineering
were most likely to be found at Research Universities," at both the

undergraduate and the graduate levels (Matyas and Malcolm, 1991). For instance, the University of Chicago has established a one-semester program in which graduate students are paired with "experienced faculty who are teaching in the school's core curriculum" (Cheney, 1990).

Career Interventions

A technique that has proven successful at the postdoctoral and professional levels is networking: groups of women meet together, with or without a facilitator, to provide encouragement and to discuss special academic, technical, and social problems that they are facing, while others suggest solutions to related problems. For instance, the American Physical Society, through its Committee on the Status of Women in Physics, is creating a network for graduate women in physics by conducting workshops on career counseling and sponsoring hospitality suites at the society's annual and regional meetings.

Several researchers have reported that marriage and motherhood do not have consistently negative effects on the employment status, publication rates, and salaries of women scientists and engineers (Zuckerman, 1987). Nonetheless, recent data imply that assistance in handling family responsibilities, perhaps through assistance with child care, can help to remove at least one barrier to full-time employment (Women's Bureau, 1988). Many employers have implemented a variety of programs—such as on-site child care centers, voucher systems to subsidize child care costs, programs for part-time or emergency care, job sharing, voluntary reduced time, flextime, and work-at-home options—in order to retain their female employees. A 1984 survey by the National Employer-Supported Child Care

Project found that of the 178 responding companies, 90 percent said that such child care services had improved employee morale and 85 percent said that their ability to recruit had been affected positively.

Academe

A report from the NRC's Committee on the Education and Employment of Women in Science and Engineering, *Climbing the Academic Ladder: Doctoral Women Scientists in Academe* (CEEWISE, 1979), explored the status of women in faculty, postdoctoral, and advisory posts, finding that women scientists were (1) concentrated in the lower ranks and in off-ladder positions, (2) typically paid less than their male colleagues at the same rank, and (3) less likely than men to be awarded tenure. A subsequent report, *Career Outcomes in a Matched Sample of Men and Women Ph.D.s: An Analytical Report* (CEEWISE, 1981), indicated that (1) these differences remain even when men and women are closely matched by education, experience, and type of employment and (2) the disparities in pay and advancement are not explained by what are traditionally considered important factors—the perceived greater restraints on career mobility or greater likelihood that women have in the past interrupted their careers for child-rearing. In still another study, CEEWISE (1983) found that between 1977 and 1981:

- In the major research universities, women held 24 percent of the assistant professorships, but only 3 percent of the full professorships.
- Women scientists were still twice or three times as likely as men to

hold nonfaculty (instructor/lecturer) appointments, an increase since 1977.

- In general, recent women Ph.D.s were found in junior faculty positions in proportions exceeding their availability in the doctoral pool.
- Promotions of junior faculty showed wide sex differences: in the group of top 50 institutions (ranked by R&D expenditures), for example, three-fourths of the men, but only one-half of the women, were promoted from assistant professor to a higher rank in those years.
- Overall, the proportion of women scientists who were tenured continued to be lower than for men.
- After controlling for rank, salary differences for men and women persist in most fields, especially in chemistry and the medical sciences.

Although data in the Doctorate Records File and from the American Council on Education, the National Center for Education Statistics, and several professional societies reveal that while the proportion of doctorates earned by women across almost all scientific fields has risen dramatically in the last decade (in general, women receive about 27 percent of Ph.D.s each year), their employment on faculties has not. For instance, the number of women Ph.D.s in physics has increased by a factor of two in 10 years, but their rate of employment on physics faculties has shown little change. Similarly, the number of women on S&E faculties is small in engineering (in which the number of women Ph.D.s increased by a factor of five) and in chemistry (where women are also getting Ph.D.s in very large numbers). On the other hand, women are quite successful in gaining faculty positions in biology.

Granted that faculty size in general has not been growing very much, the fraction of women hired and promoted still appears to be below what might be expected on a proportional basis. The number of women faculty increased in each academic rank between 1972 and 1985 (NSF, 1988). However, they made up only 28 percent of the total full-time instructional faculty in 1985 and were clustered (55 percent) in the lower ranks of assistant professors or instructors in contrast to 70 percent of all men, who were mostly professors or associate professors. By discipline the lowest percentages of tenured women faculty were in the physical sciences, mathematics, and the environmental sciences in 1987. Nonetheless, the prospects for employment in academe may be better for women today than in 1977 and 1981. At the same time that the numbers of women leaving graduate school are increasing, the availability of positions in colleges and universities, where most women scientists have historically been employed, is expected to be greater than in the past because of retirements by many current faculty.

Between 1974 and 1980 NSF, through the National Science Board, established innovative targeted programs for faculty—funding short courses, the Female Visitation Science Program, science career workshops, and the Dissertation Completion and Junior Female Scholar Awards. More recently, NSF programs for S&E faculty fall within the purview of its Directorate for Education and Human Resources (EHR) and include visiting professorships for women and 100 Faculty Awards to assist tenured but not yet full professors in their research for a five-year period. About 82 percent of NSF's 1991 budget is otherwise devoted to the regular research programs and is available to male and female scientists.

While intervention models targeting women faculty are few in number, women faculty at some institutions have been able to effect change by such means as establishing their own informal network. For instance, women faculty in science and engineering at the University of Michigan-Ann Arbor organized themselves as Michigan Women in the Sciences, initially to provide moral support for each other. Less than a decade later, the group had worked with the administration to establish a Women in Science Program that assists both faculty and students (precollege and undergraduate) through such activities as publishing a Resource Directory of Michigan Women in Science, a speakers' bureau, and a computer conference system for women engineering graduate students (Davis, 1990).

Industry

Industrial employers have cautiously implemented a large number of programs to attract and retain women scientists and engineers. In addition to starting salaries comparable to, and sometimes higher than, those offered to men, the private sector has also responded to the potential conflict between "the previously separate worlds of work and family" (Nieva, 1985). Merton (1973) has described the pressures scientists feel to be creative and to stay abreast of developments in their field. Time and scheduling problems are major sources of work-family conflicts (Nieva, 1985; Pleck, Staines, and Long, 1978; Galinsky, 1988) and can affect one's productivity (Galinsky and Stein, 1989). Consequently, some companies have implemented policies to promote a better balance between career and family responsibilities (General Mills, 1981; Reskin and Hartmann, 1986;

Belsky, 1989).[6] To retain their S&E employees, other companies offer flexible benefits packages that include health insurance, dental and/or vision coverage, life insurance, pension plans, and a reimbursable (pretax) employees' spending account. Another means of retaining employees is the availability of an education package. Tuition-reimbursement programs may enable an employee to develop or expand his or her knowledge in a work-related discipline. In addition, some companies provide similar benefits to the employee's spouse and child.

As the labor supply tightens, the current employment situation for women may change for the better, particularly if effective intervention programs are widely disseminated. For both campus and company, it might be useful to sensitize managers to situations of inequity by discussion, films, etc. For instance, Hewlett-Packard has established a program to increase the sensitivity of managers, to teach "them about their own gender biases and about different cultures and races and to inform them about the company's needs in terms of employee training and development" (Catalyst, 1988). Mentoring is also important for enhancing the performance of employees, especially women employees. Well-thought-out mentoring guidelines are available (e.g., Corning Incorporated, 1990; Catalyst, 1990),

---

[6] Such policies have resulted in a smorgasbord of programs: job-sharing, in which either spouse may share a job outside the home with another employee, perhaps even his or her spouse; part-time work, often with full-time benefits; assistance in locating, obtaining, and improving the quality of child care; reimbursement or direct provision of child-care services; flextime, which allows an employee to work a full day whose starting time varies within a stated (usually two-hour) period; elder care, "providing some type of assistance with the daily living activities for an elderly relative who is chronically frail, ill, or disabled" (Galinsky and Stein, 1989); parental leave; and alternative work schedules for full-time employees, not restricted by an eight-hour day, five days each week.

and in some companies, each new woman employee is able to choose a mentor from a list of suitable volunteers. Another approach that has been found helpful in some companies is building in accountability by basing some fraction of the performance appraisal of the manager on affirmative action performance—specifically, hiring, promotion, and development of female employees (Latham and Wesley, 1981; Catalyst, 1991a). Tenneco, for example, "[bases] a significant portion of each division's executive bonus pool on whether that division meets all of the stated goals for hiring and promoting women and minorities" (Catalyst, 1991b). Such accountability approaches might also have applicability to colleges and universities.

In general, businesses are responding to the issues of "company climate" by including women in on-the-job and inservice training exercises that provide additional skills to women for various technical and managerial positions. Still other private companies, such as Corning, Tenneco, and Xerox, have established women's forums and provided opportunities for networking among women employees (McKee, 1991). Among effective interventions are assistance with child and elder care; flexible work arrangements including flextime, flexplace, and job sharing; and training for supervisors in techniques for making jobs "doable" for women and for effectively assessing women's performance in those jobs. Nevertheless, many of these interventions are single efforts that produce only limited results. In many cases, women still find themselves isolated, with few women in senior positions to serve as role models and/or colleagues.

Government

Some programs authorized by the U.S. Office of Personnel Management (OPM) for application throughout the Civil Service might aid in attracting and retaining women scientists and engineers. For instance, OPM (1988) believes that generous leave, health benefits, flexible and compressed work schedules, leave for parental and family responsibilities, part-time employment, job sharing, and leave transfer programs are enticements for continued employment within the federal sector. In addition, the President's Council on Management Improvement implemented "flexiplace" on February 12, 1990, to enable federal employees to work at home; although still considered a pilot project, the 27 participating federal agencies cite benefits including increased worker productivity and decreased overhead costs, absenteeism, and turnover (Segal, 1991). However, many of these benefits are not accessible to federal scientists and engineers, male or female, whose major work activities are primarily research and development (R&D, 24.8 percent), design (8.9 percent), data collection and processing (7.9 percent), natural resource operations (7.9 percent), and management (5.9 percent) (NSF, 1989).

Programs targeted particularly to federally employed women scientists and engineers are often initiated by individual federal agencies. For example, 55 percent of the employees of the U.S. Environmental Protection Agency (EPA) are women and minorities: women, including minority women, comprise 48 percent of its total work force and 25 percent of its S&E work force (Reilly, 1990). However, most women and minorities at EPA are employed in nonsupervisory positions, at GS-10 or below, a situation that must be reexamined in consideration of the changing U.S. demography. To address this and other issues, Administrator William K. Reilly established EPA's Study on Cultural Diversity in its Work force, whose findings and recommendations are expected in November 1991.

The U.S. Department of Energy's programs for retention of women scientists and engineers include on-site child care facilities, wellness programs (e.g., on-site mammograms), employee assistance programs (counseling and referral for problems related to family, drugs, alcohol, and stress), parental leave programs, informal flexible schedules with the option of part-time employment, and summer recreational programs for dependent children, as well as programs such as Invitation to Lead, Management Development, Executive Development, and Leadership Development for Senior Management (not all laboratories offer all programs). Participants in a meeting convened by DOE's Office of University and Science Education Programs concluded that "there must be effective support for programs which are designed to increase upward mobility of women." Their recommendations include the following:

- release time for women scientists to engage in outreach activities to educate and excite students to the career opportunities available to women with technical backgrounds;
- policies for training and retraining including flextime, educational leaves of absence, and experience as a detailee to HQ and/or operations;
- women's committees to encourage communication and support professional development and mentoring for young people;
- vigorous efforts to examine pay equity, ranking, promotion, and recruitment of women and to make the findings public; and
- corrective action to address the current concentration of R&D women in the lower ranks of Laboratory S&E staffs (DOE, 1991).

Priority Issues

The array of programs that have been mounted over the years to encourage women to enter science and engineering careers—and to remain there—makes it difficult to identify and promote particularly effective intervention models. The Committee has selected "intervention programs" as the subject of its first annual conference in 1991. The conference is the first of several steps that might be taken to enhance current intervention models. The conference has as its goals:

- reviewing the spectrum of postsecondary programs supported by the federal government and the private sector to increase the number of potential and practicing scientists and engineers;
- delineating effective components of programs that increase the number and quality of U.S. scientists and engineers, with particular reference to women; and
- developing models that can be duplicated by participants.

Priority must also be given to:

- collecting and disseminating systematic data on programs at the undergraduate and graduate levels to sustain the flow of talented women into S&E careers and
- encouraging the development of reliable "outcome measures" to assess the specific contribution of program components to career outcomes for women.

Questions for the Committee to consider during the next three years include:

- Is intervention good public policy?
- What evidence shows that intervention programs work?
- Are intervention programs more successful at the regional level, when an entire organization is involved, or when directed at a single department (either in industry or academe)?
- What effective models are available at the undergraduate, graduate, and career levels?
- Which ones target specific groups, disciplines, and issues?

4

CAREER PATTERNS

Upon completion of an advanced degree in a science or engineering discipline, an individual may choose to pursue either direct employment in the technical work force or postdoctoral study. In many institutions, postdoctoral appointees have an ill-defined status, and all too often nobody seems to be responsible for their progress and prospects. At the same time, employment of women scientists and engineers in the United States is not as high as for men. In general, women scientists and engineers are more likely to be underemployed or underutilized than their male counterparts. As noted earlier, they are promoted more slowly. In academe they are more likely to be in the lower faculty ranks and less likely to be tenured. Proportionally fewer women than men are employed in industry, and those few are less likely to attain management positions. Women's salaries are lower within the same rank, even when allowance is made for their being younger. Among the measures that have been found helpful to advancing the careers of women scientists and engineers are (1) the availability of female role models and mentors among the more senior S&E work force and (2) access to support networks. These statements have been well-documented, as will be discussed in this chapter. In summary, we believe a greater understanding is necessary of the unique problems encountered by women as they attempt to establish themselves as practicing scientists or engineers.

Postdoctoral Appointments

Postdoctoral appointments are special because they bridge the gap between the Ph.D. and employment. A period (1-3 years) of postdoctoral research is regarded as almost a prerequisite for faculty positions in leading research departments in the physical sciences, the life sciences, and certain areas of engineering. As shown in Table 15, the percentage of U.S. citizens planning to accept postdoctoral positions prior to employment increased significantly between 1970 and 1985 but has remained relatively stable during the past five years.

Postdoctoral appointments are typically of two kinds: a true fellowship, often portable and with no duties attached; and a research appointment, which represents an extension of the standard graduate research assistantship position. Postdoctoral appointments are often prestigious and are sometimes viewed as a sort of glorious interlude between the vicissitudes of graduate study and the responsibilities of permanent employment, a period when a young scientist can enlarge and diversify her/his knowledge and capabilities without external pressure. According to Frank Press, president of the National Academy of Sciences,

> Any beginning researcher who has not worked closely with an experienced scientist is missing one of the most important aspects of a scientific education. Similarly, any experienced researcher who does not pass on to younger scientists a sense of the methods and norms of science is significantly diminishing his or her contribution to the field's progress (NAS, 1989).

However, some researchers report that is increasingly difficult to find qualified young people to work as postdoctoral fellows, because fewer bright young Ph.D.s are going into science (Simon, 1991). The low compensation for university postdoctoral positions represents a common problem, although salaries for some of the special postdoctoral fellowships

TABLE 15: Postgraduation Plans of Science and Engineering Doctorates (U.S. citizens only), 1985-1989

	With def plans (%)	Total (N)	Postdoc (%)	Empl (%)
1970				
Total, S&E	79.9	10765	21.4	78.6
Total, Sciences	80.0	8901	24.9	75.1
Total, Engng	79.4	1864	4.5	95.1
1985				
Total, S&E	72.6	8716	34.7	65.3
Total, Sciences	72.2	7808	37.8	62.2
Total, Engng	76.4	908	7.8	92.2
1986				
Total, S&E	74.3	8810	35.7	64.3
Total, Sciences	74.2	7853	38.9	61.1
Total, Engng	74.6	957	9.5	90.5
1987				
Total, S&E	73.4	8633	37.8	62.2
Total, Sciences	73.4	7575	41.5	58.5
Total, Engng	73.6	1058	11.3	88.7
1988				
Total, S&E	74.7	9097	38.1	61.9
Total, Sciences	75.0	7888	42.3	57.7
Total, Engng	73.0	1209	10.6	89.4
1989				
Total, S&E	75.8	9070	37.0	63.0
Total, Sciences	76.2	7833	41.2	58.8
Total, Engng	73.1	1237	10.7	89.3

SOURCE: Susan T. Hill, *Science and Engineering Doctorates: 1960-89* (NSF 90-320), Washington, D.C.: National Science Foundation, 1990.

at national and industrial laboratories are nearly competitive with starting salaries in more permanent positions (NSF, 1990b; OSEP, unpublished data). Since the average age at doctorate completion now ranges from about 29 to 32, depending on field (Tuckman et al., 1990), a new science or engineering Ph.D. of either sex is likely to face family responsibilities, so that salary is an important consideration. Although many sponsors have made efforts to keep stipends competitive, financial problems remain. Zumeta (1985) found that the income loss sustained during a postdoctoral appointment was unlikely to be recovered later, relative to a new Ph.D. who entered employment directly.

The National Research Council's 1981 report on *Postdoctoral Appointments and Disappointments* looked at the issue from a slightly different perspective. It found that married men were less likely than single men to accept postdoctoral appointments, but that the reverse was true for women. This suggests that many men who must support families do not feel that they can sacrifice the additional income of a permanent position. However, women whose mobility is constrained by factors such as marriage and child-rearing may welcome postdoctoral opportunities as opposed to permanent positions, even when prolonged beyond the usual two-year period, as a viable alternative to leaving the profession. It appears that men are less likely to face such a choice. Nonetheless, despite various potential drawbacks, postdoctoral appointments remain a useful option for the increasingly frequent problem of dual-career couples seeking appointments in the same location or institution. This may explain the fact that, in the past, women were about 50 percent more likely to go into and remain in postdoctoral appointments than men.

In the past, there were substantial differences by sex in the rates of seeking as well as finding postdoctoral appointments, but the most recent data suggest a virtual disappearance of these differences (Coyle, 1986). Hargens (1971) and Reskin (1976) found that postdoctoral fellowships are awarded to men more often and that men had far better career outcomes than women with similar training; but in that study, data from the late 1960s were used and the numbers of women scientists available for study were relatively small. The situation has changed in many respects since then: OSEP doctoral data show that there are now no significant sex differences in the frequency and duration of postdoctoral awards. Fields such as engineering, which have enjoyed high demand for some years, routinely show little or no sex difference in placements. On the other hand, information on the size of stipends is not available, and sex differences may exist there.

Still another issue is whether postdoctoral appointments contribute effectively to the career development of new Ph.D. recipients. Some people question whether completion of a postdoctoral appointment facilitates getting a good starting faculty position, but this is clearly field related. The 1981 NRC study found that many postdocs, "like planes waiting to land, were stacked in a holding pattern" because there were not enough available jobs. This basic conclusion was confirmed four years later (Zumeta, 1985). As shown in Table 15, the percentage of Ph.D.s having definite plans for employment, versus postdoctoral appointments, has averaged about 60 percent in the sciences and 90 percent in engineering during the 1985-1989 period. This contrasts with the 75 percent and 95 percent reported for science and engineering, respectively, in the 1970-1975 period. Some individuals seek postdoctoral appointments when employment demand is low, which may be the case for the most recent period. However, the market seems not yet to have adjusted, since some research faculty report that their postdoctoral fellows are having a difficult time in finding employment. As the number of unfilled openings in industry and academe increases, however, the problem of postdoctoral placement in permanent positions should largely disappear.

Employment in a Scientific or Engineering Field

Current and projected levels of labor force participation indicate that women will continue to make up a significant portion of the U.S. work force. At the same time, women's position in the labor market, when measured by earnings, is relatively disadvantaged compared to men's. The increase in women's economic activity over recent decades, concomi- tant changes in household and family structure, and women's continued unequal status in the S&E labor force (as will be further documented) can raise crucial questions about many of the assumptions that underlie prevailing employment policies and related policies in such areas as the family and education. These policies, and their implications for both female and male scientists and engineers, need to be studied in light of the recent changes in women's participation in the labor market and in family structure, in order to determine whether they are meeting their aims and even whether their aims are still appropriate. Efforts to do so are already under way in the federal government and in a variety of organizations such as private foundations and women's groups. In fact, many changes in employment policies are beginning to be implemented, but these changes must be monitored.

The problem is not with getting women into S&E careers, but with helping them move up in those careers (Knapp, 1983). In all employment

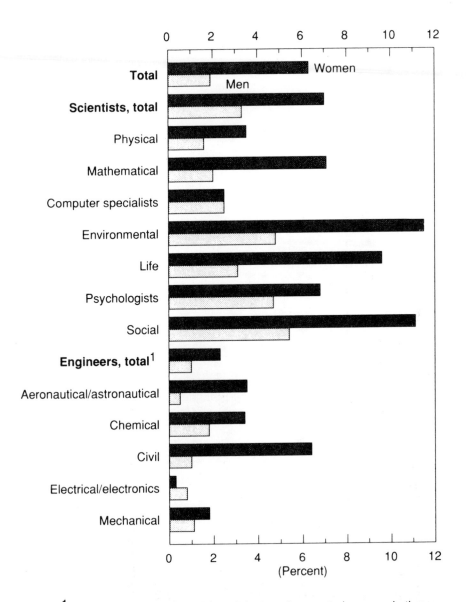

¹Includes industrial, materials, mining, nuclear, petroleum, and other.

SOURCE: National Science Foundation, Women and Minorities in Science and Engineering (NSF 90-301), Washington, D.C.: NSF, 1990.

Figure 8. S/E underemployment rates of men and women, by field, 1986.

TABLE 16: Reasons Given for Not Being Employed Full-Time by Science and Engineering Doctorate Recipients, 1989

Reason Employed Part-Time	Year	Total	Male Number	Percent*	Female Number	Percent*
TOTAL	1989	21,562	12,727	59.0	9,135	41.0
Prefer Part-Time Employment	1989	8,202	5,645	44.4	2,557	28.0
Full-Time Position Not Available	1989	3,931	2,397	18.8	1,534	16.8
Family Responsibilities	1989	4,058	443	3.5	3,615	39.6
Other	1989	3,287	2,592	20.4	695	7.6
Combined Pref./ Family Respons.	1989	409	65	5.1	344	3.8
Combined Full-Time Not Avail./Family Respons.	1989	90	0	0.0	90	1.0
No Report	1989	1,885	1,585	12.5	300	3.3

Not Seeking Employment

TOTAL	1989	4,202	1,509	35.9	2,693	64.1
Health, Personal Reasons (Temp)	1989	378	148	9.8	230	8.5
Family Responsibilities	1989	1,477	50	3.0	1,427	53.0
Suitable Job Not Available	1989	353	139	9.2	214	7.9
Other	1989	929	574	38.0	355	13.2
Combined Health, Personal/ Family Responsibilities	1989	36	0	0.0	36	1.3
Combined Family Respons./ Job Not Available	1989	54	0	0.0	54	2.0
Student	1989	861	571	37.8	290	10.8
No Report	1989	114	27	1.8	87	3.6

*Percentages for all but the first row represent those, of the number (by sex), citing the given reasons. For instance, the 5,046 men preferring part-time employment in 1987 comprise 46.7 percent of the 10,812 male S&E Ph.D.s surveyed in 1987.
SOURCE: Office of Scientific and Engineering Personnel, Survey of Doctorate Recipients.

sectors, women are promoted more slowly and paid less than comparable men (Vetter, 1989b; NSF, 1990b). Furthermore, over all employment sectors, more than three times as many women as men with S&E degrees are **underemployed**—defined as working not at all or in part-time positions (Figure 8). In 1986 the underemployment rates for men and women scientists and engineers were 1.9 percent and 6.3 percent, respectively—3.3 percent and 7.0 percent, respectively, in science and 1.0 percent and 2.3 percent, respectively, in engineering (NSF, 1990b).

Interestingly, labor force participation rates vary for women among S&E fields (Thurgood and Weinman, 1990; NSF, 1988). They are especially well utilized, for instance, in electrical/electronics engineering and computer science, which are large employment fields, but less well utilized in aeronautical/astronautical, chemical, civil, and mechanical engineering and in the mathematical, environmental, life, and social sciences. In general, the positions occupied by female scientists and engineers are not those of power and prestige or those that permit them to engage in policy making or consulting, the activities that provide the greatest incentives and give the individual the greatest visibility outside his or her own institution (see, for instance, Woodward, 1990). In addition, women more often than men report that they are **unemployed** or working part-time because of family responsibilities (Table 16).

Academe

Almost two-thirds of the women Ph.D.s on S&E faculties, as compared to about 40 percent of the men, either were not tenured or were

TABLE 17: Tenure Status of All U.S. Doctorate Recipients in Science and Engineering, 1989

Tenure Status	Year	Total	Male		Female	
			Number	Percent	Number	Percent
TOTAL	1989	221,784	181,541	82.0	40,243	18.0
Tenured	1989	121,986	107,473	88.0	14,513	12.0
Year Granted						
1949 or earlier	1989	15	15	100.0	0	0.0
1950-1954	1989	321	321	100.0	0	0.0
1955-1959	1989	1,733	1,673	97.0	60	3.0
1960-1964	1989	6,271	6,078	97.0	193	3.0
1965-1969	1989	14,145	13,430	95.0	715	5.0
1970-1974	1989	23,398	21,510	92.0	1,888	8.0
1975-1979	1989	25,298	22,530	89.1	2,768	10.9
1980-1984	1989	23,784	19,861	83.5	3,923	16.5
1985-1989	na	na	na	na	na	na
Year Not Reported	1989	3,722	3,226	86.7	496	13.3
Not Tenured	1989	64,131	45,645	71.2	18,486	28.8
Postdoc Appt	1989	11,892	8,491	71.4	3,401	28.6
In Track	1989	33,470	24,834	74.2	8,636	25.8
Not In Track	1989	18,769	12,320	65.6	6,449	34.4
Tenure Not Applicable	1989	16,178	11,765	72.7	4,413	27.3
No Report	1989	19,489	16,658	85.5	2,831	14.5

SOURCE: Office of Scientific and Engineering Personnel, Survey of Doctorate Recipients.

holding positions where tenure was not applicable (Table 17). To some extent, these numbers reflect the recent increases in women Ph.D.s in these fields, though significant differences in rank, income, and tenure status remain after this factor is considered. According to one study, the lack of change within institutional structures has allowed sex inequities in the hiring, promotion, and tenuring of faculty to continue substantially unchanged (ACE, 1988). This is supported by the results of institutional self-assessments such as that recently completed by the Faculty of Arts and Sciences' Standing Committee on the Status of Women at Harvard University:

> . . . while most no longer have to contend with blatant discrimination, aspiring women scientists still face barriers to advancement that their male colleagues do not. . . .
> "Many cases don't involve people of ill will, just people doing what they've always done and not realizing that the way in which they operated does shut certain people out," says Barbara J. Grosz, McKay professor of computer science, who chairs the committee.
> Such tacit discouragement can involve anything from using "macho" stereotypes when describing research style to assigning traditionally female responsibilities—such as entertaining outside speakers and prospective faculty—to women members of a department. . . . In some departments, verbal and gender-specific harassment (e.g., discouraging women from pursuing scientific careers simply because they were women) are prevalent even though sexual harassment may not be. . . . the fewer women a department has to begin with, the harder it is to attract and retain others. With so few women in senior positions, women graduate students and junior faculty often feel lonely and isolated, unable to make the personal connections that successful careers often depend on, or to look to colleagues for support. The solution, says Grosz, is not to hire a few token role models but to achieve "critical mass" for women in every department. "Women need to know that it's a normal thing to be a scientist," she says (Harvard University, 1991).

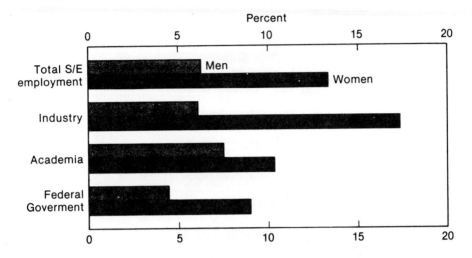

SOURCE: National Science Foundation, Women and Minorities i Science and Engineering (NSF 88-301), Washington, D.C.: NSF, 1988.

Figure 9. Average annual employment growth, by sector of employment and sex: 1976-1986.

Industry

The position of women in industry may be described in similar terms to their position in academe, although their rate of employment in this sector has improved substantially since equal employment opportunity began to be implemented and monitored (Figure 9). In recent years, women's employment in science and engineering showed the greatest gain in the industrial sector (17 percent), with increases above 15 percent in the areas of computer science and engineering (particularly aeronautical/ astronautical, chemical, and electrical/electronics), as shown in Figure 10. By comparison, employment in industry by male scientists and engineers rose only by about 6 percent.

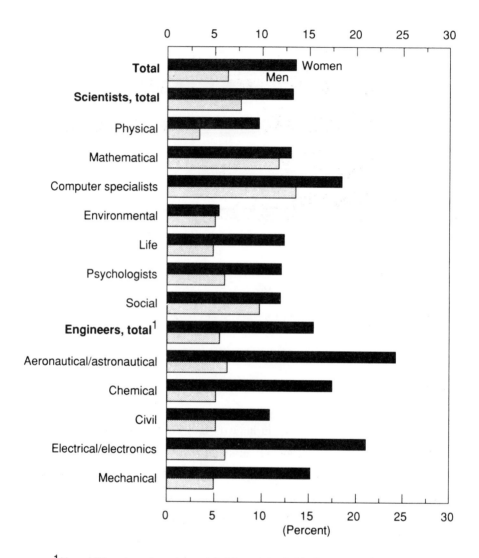

¹No additional engineering subfields are available for 1978.

SOURCE: National Science Foundation, Women and Minorities in Science and Engineering (NSF 90-301), Washington, D.C.: NSF, 1990.

Figure 10. Average annual employment growth rate of scientists and engineers, by field and sex, 1978-1988.

Micro- and macro-inequalities also exist in industry, although not necessarily identical to those in education and academic employment in the sciences and engineering. A blatant example appeared in *The Wall Street Journal*: when the retiring chairman of a major U.S. company was asked, "Is there much of an opportunity for women in the steel industry?," his response was,

> I can see a day when a lot of the top administrative positions—like treasury, law, accounting, and public affairs—could clearly be handled very well with women. It is tougher for us to recruit women into manufacturing jobs. There are fewer engineers who are women. We don't start at 8 and go home at 5. Our manufacturing process goes 24 hours a day, 7 days a week. They're coming to work when it's dark, going home when it's dark. Let's face it, a lot of ladies don't find it too appealing (Schellhardt and Hymowitz, 1989).

The inequalities are usually more subtle. Reskin and Hartmann (1986) described inequities common to the workplace in general:

- relying on informal referrals ("old boy network") rather than advertising directly to fill vacancies;
- using that same network rather than standard performance evaluations when considering individual employees for promotions;
- "precluding spouses from working in the same department of company;"
- linking promotions to one's length of employment;
- condoning coworkers' remarks and actions "calculated to put [women] in their place by emphasizing their deviant gender status . . . [in] occupations that have been defined as male;" and
- excluding women from "informal networks in which information is shared and alliances develop."

In industry, the existence of such inequities contributes to a "company climate" that is often not as supportive of women employees as it could be.

It is well documented that women are much less likely to attain management positions than men (NRC, 1980; Dipboye, 1987; NSF, 1990b). This could be particularly discouraging for women scientists and engineers in light of NSF projections:

> Over the 1988-2000 period, the number of jobs for scientists and engineers in private industry is expected to increase to over 600,000, with three-fifths of these new jobs to be located in the services-producing sector. . . . Between 1988 and 2000, the occupational composition of industry jobs is expected to change away from production and assembly-line jobs toward professional, *managerial*, and technical occupations (National Science Board, 1989).

Government

During the 1990 Workshop on Recruitment, Retention, and Utilization of Federal Scientists and Engineers at the National Academy of Sciences, John M. Palguta, deputy director of the U.S. Merit Systems Protection Board (MSPB) Office of Policy and Evaluation, noted that

> Efforts to manage the federal work force today must operate in an environment that is significantly different from 20 years ago. . . . Shifting demographics, the rapidly changing international climate, and the declining image of federal employment all argue for some fundamental shifts in the way federal personnel management is carried out. This may be especially true as the government struggles to recruit, motivate, and retain a large cadre of well-qualified engineers and scientists (Palguta, 1990).

The White House Task Force on Women, Minorities, and the
Handicapped in Science and Technology (1989), after studying 236,757
scientists and engineers employed full-time by the federal government in
1987, found that while employment of women and minorities had almost
doubled during the previous decade, only 10 percent of the employed S&E
Ph.D.s were women and, of that group, only 57 percent (compared to 75
percent of the same cohort of white males) earned $50,000 or more a year.

Data acquired by the Commission on Professionals in Science and
Technology (Vetter, 1991) show that the participation of women scientists
and engineers in the federal work force varies by occupational title from a
low of 3.0 percent in agronomy, 3.8 percent in mining engineering, and 4.8
percent in metallurgy to highs of 50.5 percent in sociology, 44.1 percent in
social science, and 41.6 percent in botany. Racial/ethnic minorities
comprised 14.6 percent of the 263,892 federal S&E employees on September 30, 1990. The greatest numbers of minority women were found in
chemistry, social science, general engineering, and general biological
science. Minority women were least represented in ceramic engineering
(3.0 percent), the biological science fields of ecology (3.5 percent) and
botany (5.6 percent), sociology (3.9 percent), and the physical sciences of
astronomy/space science (4.5 percent) and hydrology (5.4 percent).
Furthermore, it was recently determined that only 5 of the 66 top
government positions in science are held by women:

- Administrator, National Telecommunications and Information
  Administration;
- Surgeon General, Public Health Service;
- Commissioner, Bureau of Labor Statistics;
- Chairman, Consumer Products Safety Commission; and

- Administrator for Pesticides and Toxic Substances, Environmental Protection Agency (Clemmit, 1990).

Though their numbers are small, one would expect federally employed women scientists and engineers to earn salaries equivalent to those of their male counterparts. However,

> In federal employment, women earn only 65 percent of male wages. . . . When jobs are compared based on skill, responsibility, effort, and working conditions, pay for female jobs [the author hopes that this is defined as those jobs in which women are actually employed, as opposed to those traditionally "assigned" to women] falls far short (AFL-CIO, 1990).

This confirms earlier reports that, in general, women employed by federal agencies earn only 79.2 percent of the average salary earned by males in chemistry, 78.7 percent in microbiology, and 73.9 percent in civil engineering (Babco, 1985, in Vetter, 1987).

The White House Task Force (1988) had recommended that the government tackle the potential problem of an inadequate S&E work force head-on by:

- using federal R&D programs to effect "a more diverse, world-class science and engineering work force;"
- collecting and maintaining data in order to evaluate the participation of women, minorities, and disabled persons in those programs;
- hiring and promoting scientists and engineers who are members of underparticipating groups; and
- providing dependent care services to all employees.

One of the government's first responses was examination by the Federal Coordinating Council on Science, Engineering, and Technology (FCCSET) of the issue of underparticipating groups. FCCSET's Committee on Education and Human Resources has determined that federal agencies should allocate funds designed specifically to retain women, minorities, and the disabled within the science education pipeline.

Priority Issues

Perhaps the most critical issue confronting the U.S. economy today is the diversification of the work force. Opportunities are growing for women in a wide variety of R&D and S&E positions. CWSE has therefore given high priority to developing a program of activities that will facilitate positive changes in this area. Among the issues that the Committee has determined to be priorities are the following:

- Postdoctoral appointments—
  * Are there currently differences in the postdoctoral appointments of men and women with regard to (1) size of stipend, (2) numbers of years in postdoctoral status, and (3) opportunities to work with first-class researchers in their field?
- Employment policies—
  * CWSE will seek to integrate current efforts to examine critically employment policies. Are the policies still appropriate in light of recent changes in the number of women working and in family structure? Results of such efforts will be disseminated.
  * CWSE will collect and disseminate information about the

many changes that are already taking place in employment policies——for example, employer-sponsored women's support groups, employer-sponsored women's councils designed to suggest improvements in working conditions for women, changes in nepotism rules, availability of choice of benefits (cafeteria-style benefits), changes in maternity and adoption leave policies, employer-provided or sponsored dependent care, prorated benefits for part-time employees, changing tenure and promotion policies to reflect women's extra family responsibilities, and extended possibilities for part-time and flex-time employment.

* CWSE will also undertake inquiries into less well studied areas of employment policy relevant to women in order to highlight major issues, develop research agendas, and suggest possible courses of action and their probable implications.

- Broader issues——
  * barriers to the advancement of women scientists and engineers (the "glass ceiling");
  * reasons women scientists and engineers are less represented in industry than are men;
  * the importance of mentoring and role models;
  * retention patterns for women; and
  * particular problems in the work environment for women at the bachelor's and master's levels.

The Committee on Women in Science and Engineering would seek to integrate these efforts and to underscore the importance of the issues involved to policy makers and academic researchers.

5

MEASUREMENT FOR SCIENTIFIC
AND ENGINEERING HUMAN RESOURCES

Changes in U.S. demography, in the infrastructure of science and
engineering education, and in the career paths of scientists and engineers
are adequate to begin to understand the situation of women in science and
engineering. However, as noted in OSEP's 1991 Strategic Plan, "a fundamental barrier to rigorous analysis of the issues raised in this plan is the
lack of adequate data." Much remains to be learned, particularly
concerning the role of women in the scientific and engineering (S&E)
human resource base.

Education Infrastructure

As noted in Chapter 2, both formal and informal mechanisms
encourage females to pursue the education they need to become practicing
scientists and engineers. Data and analysis of their effectiveness are both
inadequate, but existing data do show that:

- The entry of women into higher education, in general, grew by 41
  percent between 1970 and 1983 and accounted for 52 percent of
  1989 bachelor's degree recipients.
- Explicit recruitment and retention efforts, such as those in
  engineering, can raise the number of female graduates in all fields
  from about 800 to 11,000 per year in a decade.
- Women showed increasing persistence through S&E doctoral
  programs.

- Compared to men, women are lost from the S&E pipeline through lower rates of recruitment and retention. These losses occur at different rates and at different places in the pipeline, depending on field, which suggests that completion is more directly related to field-specific characteristics than to commonly postulated factors such as marriage and parenthood.

A large body of data and interpretation already exists on under-graduate and graduate enrollments and on attainment rates for women in science and engineering at all degree levels. However, recruitment activities of academic administrators and S&E faculties indicate that "colleges and universities have no systematic ways of obtaining this much-needed information" (Keith and Keith, 1990). The Committee did learn that some institutions, such as University of California-Berkeley (1989), do collect and provide such data in internal reports. Furthermore, except for the samples included in NLS-72 and High School and Beyond (Burkheimer and Novak, 1981; Eagle et al., 1988), longitudinal data are not currently collected on continuation rates by field from undergraduate degree comple-tion to graduate enrollment, although such data are necessary in order to create a supportive campus climate and devise appropriate intervention strategies. Overall comparisons for Ph.D. completion rates based on undergraduate major do show that the sex differences in degree attainment in many fields are quite small: this suggests that those fields that exhibit significant losses of women from the Ph.D. pipeline are a departure from the norm. Nonetheless, many researchers feel that women are more likely than men to pursue master's degrees rather than Ph.D.s; the available data are not sufficiently disaggregated to confirm or deny this.

Similarly, data are not collected in a comprehensive data base to show the comparative success of women in obtaining financial support for studies in the sciences and engineering. Data extracted from such sources as the Doctorate Records File do indicate financial success for specified segments of the S&E human resource base, but in general, we do not know how the success rate compares to application rate or the degree to which women, compared with men, apply for various forms of financial assistance. Instead, as was done for the NSF Graduate and Minority Graduate Fellowship Programs (see Chapter 2), one must pull such data on a program-by-program basis. Analyzing such data is essential to verify that:

- women are more likely than men to discontinue their education because of insufficient funds; and
- women offered financial aid at the beginning of their undergraduate education are more likely not only to earn a bachelor's degree in science or engineering, but also to be retained in the education pipeline through completion of doctoral studies.

In its role as a monitor of the progress of women in science and engineering, the Committee will examine how women fare in various financial aid programs available at the undergraduate and graduate levels of the education pipeline.

Quantitative data are also lacking on informal mechanisms within the S&E education infrastructure. Much anecdotal evidence exists, for example, on the importance of the media and parents influencing precollege girls to be interested in science and engineering, but little statistical data has resulted from the limited quantitative research on this issue. Similarly, the Committee knows of no research that examines the influence of the media on the retention of women at the undergraduate and graduate levels, although some studies have shown that having a parent, spouse, or mentor who is a scientist or engineer is influential in both recruitment and retention at those levels of the education pipeline. Nontraditional methods of research—for example, ethnographic studies and focus groups—may be appropriate vehicles for deriving initial information on the basis of which more rigorous information-gathering efforts might be mounted.

To follow up on this latter finding, assessments should be made of the effects of role models and mentors on recruitment and retention. While many institutions and professional societies assert the importance of mentoring and networking, we do not know definitively how important this is for students in the sciences and engineering—either female or male. In fact, a comprehensive search for research in the area of mentoring yielded no studies focusing solely on these disciplines; rather, most research to date has examined the mentor-protege relationship in the fields of education and business administration. The 1989 NAS report, *On Being a Scientist*, found that mentors can be quite influential on one's career advancement, but no findings from quantitative research are readily available to support or refute the anecdotal evidence submitted by eminent members of the S&E community. A more comprehensive examination of mentoring in science and engineering must be undertaken in order for this phenomenon to be understood more thoroughly.

Finally, the Committee on Women in Science and Engineering believes that more data are needed on the institutional factors affecting the recruitment and retention of women in S&E disciplines. A related finding of the AAAS study was that, "In general, there did not seem to be a central source of information or coordinated effort concerning recruitment and retention of women and/or minorities in science and engineering" (Matyas and Malcom, 1991). This can lead to negative recruitment and retention: after expressing an initial interest in science and engineering (S&E) studies, women more often than men switch to nonscience or nonengineering fields, typically basing their decisions on sociocultural and attitudinal factors rather than on their own academic talent or performance. Most data on this "chilly climate" have come from the American Association of Colleges' (AAC) Project on the Status of Education and Women, which surveyed limited populations. But because the AAC data support the reported experience of many women scientists and engineers, it is essential to have quantitative data in order to determine the pervasiveness of this negative treatment of women attempting to broaden their education in the sciences and engineering. One aspect of the campus environment requiring greater data collection and analysis is advising protocols as evaluated by female students—how are their majors and degree objectives chosen, for instance, and what degrees are chosen?

Intervention Models

Some organizations within the S&E community have taken actions to encourage more women to pursue careers in the sciences and engineering, and these interventions have had a positive effect (GUIRR, 1987). However, few of these interventions have been systematic or sustained (see, for instance, Cavanaugh, 1990); and despite the very impressive gains in recruitment and retention, the number of female scientists and engineers remains relatively small: in general, less that one out of every six scientists and engineers is a female (Table 18). The increased participation of women in science and engineering has not been enough to offset the overall

TABLE 18: Numbers of Women Employed in S&E Fields, by Race/Ethnicity: 1982-1986

| Race/Ethnicity | Women | | | | Men |
	1982	1984	1986	Percent change, 1982-1986	1986
African-American	23,000	22,900	34,500	50	80,500
Asian-American	18,900	27,000	36,300	92	190,500
Hispanic	9,500	15,200	19,600	106	73,800
Native American	1,900	1,500	2,700	42	21,000
White	339,800	452,200	608,600	79	3,581,500
All	388,900	512,600	698,600	80	3,927,800

SOURCE: Robert C. Johnson, Black female participation in quantitative domains, in Sandra Z. Keith and Philip Keith (eds.), *Proceedings of the National Conference on Women in Mathematics and the Sciences*, St. Cloud, Minn.: St. Cloud State University; compiled and calculated from National Science Foundation, *Women and Minorities in Science and Engineering*, 1982 and 1988.

trend of declining enrollments in these fields, especially in certain disciplines where projected demands are largest (see Table 1, page 8).

There have been few assessments of either participation rates for women in science and engineering or the effectiveness of programs to increase their participation. Few intervention programs have been sustained for a period long enough to permit evaluation, reap maximum benefits, and allow for institutionalization and replication of successful programs. The 1991 AAAS study (Matyas and Malcom) confirmed earlier findings that "most intervention programs have not been evaluated extensively" and added that less than half have undertaken longitudinal analyses of their programs' effectiveness. Furthermore, there has been little dissemination of materials describing interventions that have been shown to be effective (Erickson and Erickson, 1984). However, the Committee also learned that, to their credit, some individual institutions of higher education have undertaken self-assessments, which have led to programs targeted to potential women scientists and engineers. Among those of which the Committee is aware are those at the University of Maine (Schonberger, 1990), University of Minnesota-Twin Cities (Mauersberger, 1990), and Illinois State University-Normal (Jones, 1990). But the Committee has not examined those programs and can make no determination as to the appropriateness of the assessment methodologies used in the self-evaluations.

An examination of laboratory programs for women undertaken by the Department of Energy was completed recently. Participants at a November 16, 1990, meeting determined that effective programs for women scientists and engineers employed in DOE laboratories have four major characteristics:

1. ensure effective recruitment of qualified female candidates;
2. maintain strong networking and mentoring programs;
3. facilitate movement into management and senior scientist positions; and
4. encourage the expression and discussion of areas of concern (DOE, 1991).

Subsequently, the U.S. Merit Systems Protection Board (MSPB) has undertaken a special study on women in the federal government to describe their occupational and grade-level distribution, determine where there are imbalances, identify barriers to occupational mobility and promotional opportunity for women, and recommend ways to remove those barriers. According to Katherine C. Naff, project manager, this study will oversample federally employed scientists and engineers, both male and female, in an attempt to more accurately portray the status and role of women scientists and engineers in federal agencies.

Career Patterns

Data on the career patterns of women scientists and engineers are somewhat more obtainable, a necessity if policy makers are to effect changes that will increase their participation in the S&E work force. For instance, NSF regularly reports data on the employment—including unemployment, underemployment, and underutilization—of women scientists and engineers in the biennial reports, *Women and Minorities in Science and Engineering* and *Science & Engineering Indicators*. However, data are often reported 2-4 years after they were collected, giving an outdated picture of the current situation. Such data, which are used frequently to reveal trends, may delay new or changed programs designed to provide the United States with an adequate supply of qualified scientists and engineers. Thus, it is imperative that data be released to the public in a timely manner so that appropriate responses can be made by all sectors of the economy.

At the same time, greater use should be made of available data that can indicate the career paths of both female and male scientists and engineers. For instance, at a 1986 OSEP-sponsored and NSF-funded workshop, participants examined available data to gain a clearer picture of the underrepresentation and career differentials of women in science and engineering. The proceedings of that workshop, because of their analytical nature, have been used widely by policy makers in both government and industry to assess actions that might be taken to recruit and retain future women into S&E careers.

The Committee on Women in Science and Engineering plans to rely on currently available data in several studies to address issues of the employment of women in science and engineering. Among such studies are the following:

- A Study of Career Differences in a Matched Sample of Men and Women Ph.D.s in Science and Engineering: The purpose of this study is to provide data that may help measure any reduction of disparity between careers of men and women Ph.D.s during the past decade,[7] examining sex-related barriers to their participation

---

[7] This would be an update of work conducted by the earlier Committee on the Education and Employment of Women in Science and Engineering. See Nancy C. Ahern and Elizabeth L. Scott, *Career Outcomes in a Matched Sample of Men and Women Ph.D.s: An Analytical Report* (Washington,

and advancement and identifying career barriers that federal policy might help to overcome. The study of career differences will be based largely on analysis of data for a sample of Ph.D.s in OSEP's Longitudinal Work History File and Doctorate Records File and on comparisons with earlier published studies. The study will seek answers to questions such as:

* Has the distribution of employment of women by sector changed in the past decade, e.g., do universities currently employ a higher proportion of the employed women Ph.D.s? To what extent do women scientists and engineers, as compared with their male counterparts, migrate from one employment sector to another?

* Has the disparity between median starting salary and annual salary for men and women (by field, by employment sector, and by age) decreased?

* Do women in all fields still exceed men in involuntary part-time employment or unemployment immediately after receiving their degree?

* How is marital status associated with career attainment?

* What is the effect of having held a postdoctoral appointment on one's career path?

• <u>A Study of Women Scientists and Engineers Employed in Industry:</u> The reasons for the relatively low rate of industrial employment for women scientists and engineers cannot be determined from the information presently available. For example, it is not clear to what extent this difference from that of men represents problems

---

D.C.: National Academy Press, 1981).

of work location, especially in the case of two-career couples. However, current data can provide answers to other questions:

* What are the retention rates, by gender and degree level, for scientists and engineers in industry and government?
* Are women more likely to leave within five years of employment?
* What are the reasons most often cited for such exits?
* To what extent do male and female scientists and engineers pursue a career in both research and management?
* How has the percentage of women scientists and engineers in management changed during the past 10 years, and how does their number in the ranks of management compare to that of men?
* What differences exist in salary and rate of promotion for those engaged in research versus those who have moved into management, by sex?

* A Study of the Status of Women Scientists and Engineers in Academe: Since 1975, the supply of women Ph.D. scientists and engineers has grown sharply. The Committee will use data currently available from the Survey of Doctorate Recipients and Survey of Earned Doctorates to analyze such issues as the following:

* the percentage of women on science faculties, both nationwide and at leading research universities;
* promotion rates for women faculty, especially promotions from the rank of assistant professor (where most women were found in 1977) to associate professor;
* gender differences in awarding of tenure to women and

men at the same rank;
* salary differentials, analyzed separately for public and private institutions;
* off-ladder appointments (i.e., whether women continue to be represented disproportionately often in these positions, with limited opportunities for research and for improving their prospects); and
* the relationship between marital status and having children to promotion rates and salaries.

Such studies are particularly important, since inequalities between women and men in career success indicators—such as academic rank, tenure, and salary—may influence young women to seek careers in non-S&E professional fields, where they perceive less inequality. The Committee on Women in Science and Engineering believes that further progress in increasing the participation of women in science and engineering will depend heavily on making available timely and carefully analyzed data about their career status, and it will devote much effort to securing and publishing such information.

Priority Issues

To permit rigorous analysis of issues related to the recruitment, education, and employment of women in science and engineering, it will be necessary to foster the development of suitable measures. With regard to the development of **effective educational policies**, measures are needed to monitor retention of women in undergraduate and graduate programs in science and engineering—including trends in financial support for these students. To assess the effectiveness of **intervention models,** as described throughout this report, priority must be given to the development of outcome measures. With regard to issues affecting **career patterns,** finer measures of labor force adjustment are needed, including the ultimate disposition of postdoctoral personnel.

The Committee on Women in Science and Engineering plans to foster the development of these measures by facilitating the establishment of a central data-collection network and encouraging the exchange of recruitment and retention data among institutions—both those educating scientists and engineers and those employing this talented work force (colleges and universities, industry, and government agencies). The Committee has identified the following questions that must be answered in order to fulfill these objectives:

- How complete is the data base provided by the "top 25" institutions—Ph.D.-granting, B.S.-granting, HBCUs, women's colleges—with respect to financial support and degree completion rates?
- How comprehensive are the outreach, recruitment, and mentorship programs at all three degree levels and by the various employment sectors?
- How are the above programs and career development programs evaluated by institutions and employers? What use is made of such evaluations?
- How well established and evaluated are the policies on sexual harassment and programs to prevent crimes against women? How conducive to learning, with assured personal security, is the campus climate for women?

- How can CWSE promote uniform collection and reporting procedures for the data described above?

Through meetings with data-base administrators and users, as well as by undertaking its own research that relies on information collected in national and institutional data bases, the Committee proposes to answer these questions.

6

STRATEGIC PLAN
FOR INCREASING THE NUMBERS OF WOMEN
IN SCIENCE AND ENGINEERING

In his presidential address at the 125th annual meeting of the
National Academy of Sciences, Frank Press emphasized the necessity for a
cohesive science policy. Among issues that he cited as top priorities was
the need to preserve the nation's human resource pool in science and
engineering, in order to stave off potential shortages (Press, 1988). If the
goal is to encourage more American students to pursue studies in the
sciences and engineering, it will also be necessary to coordinate the many
activities required. The rationale for a concerted effort was stated
eloquently by Daniel E. Koshland, Jr. (1988):

> The threat of a serious shortage of scientific personnel
> looms in the years ahead. . . . If a shortage is a realistic
> scenario, . . . it is important to find ways to employ
> underrepresented groups more equitably—for reasons of
> national interest as well as of equality.
>   As the country expands into an ever-increasing
> technological base, the need for women and minorities in
> both academia and industry increases proportionally. It
> may cost some money, some effort, and some understand-
> ing, but the voyage to full equality can be even more
> exciting and worthwhile than the voyage into space.

However, the evidence shows the female scientist or engineer to be only a
marginal participant in the scientific and engineering (S&E) activities of the
Nation. It also reveals that the situation is only improving slowly. Not only
do female scientists and engineers represent a small proportion of the total
technical and professional population, but their abilities and training are
also often diverted from scientific activity.

105

An important strategy for ensuring an adequate supply of U.S. scientists and engineers to meet pressing national needs in an increasingly global marketplace would be to increase the representation of women from all racial and ethnic groups in S&E careers. Based on findings presented in earlier chapters of this report, the Committee on Women in Science and Engineering believes the time has now come for analysis and evaluation of earlier and current research, substantial new research, and further implementation of successful interventions.

First-Year Plan

In 1992, the Committee will emphasize the following issues as it develops its plan of action:

S&E Education Infrastructure:
- identifying educational programs that have been effective in facilitating the recruitment and retention of women in S&E careers, with emphasis on programs at the undergraduate and graduate levels of education;

Intervention Strategies/Measurement:
- encouraging the development of reliable outcome measures to assess the specific contribution of programs that enhance the flow of women into S&E careers; and
- fostering the development of finer measures of labor force adjustment, including tracking the career paths of postdoctoral personnel;

Career Patterns:

* developing a program of studies to facilitate the positive employment opportunities related to diversification in the workplace; and
* exploring issues related to the support infrastructure that makes it possible for women with family responsibilities to participate in the S&E labor force.

Long-Range Plan

The Committee has at its disposal a number of mechanisms with which to tackle these and other priority topics. The Committee may wish to:

* **stimulate research** on issues relevant to women scientists and engineers, by establishing study panels that can explore some subset of these issues in greater depth;
* **monitor the progress** of efforts to increase the participation of women in scientific and engineering careers, through workshops and conferences;
* **brief appropriate officials** on matters leading to the development of programs for women in science and engineering; and
* **disseminate current data** about the participation of women in science and engineering to broad constituencies in academe, government, industry, and professional societies, through the services available at the National Research Council.

Activities associated with each of these actions are described below.

The Committee on Women in Science and Engineering stresses that all studies of women in science and engineering should explicitly consider data on minority women, whose education and employment opportunities are both similar to and different from those of white women. Because the minority population has several subgroups, with quite different cultures, it is also important to study the differences in sociocultural and attitudinal factors among members of these groups: African-American, American Indian, Asian, and Hispanic. With that caveat, the Committee plans to undertake research on sex-specific issues:

- Recognizing the attrition that occurs at specific points in the S&E education/employment pipeline, the Committee will undertake research to determine effective modes of sponsoring and mentoring women scientists and engineers at all levels, from the undergraduate level through the postdoctoral level and in the professions. Although studies of these phenomena in the fields of education and business administration abound, little more than anecdotal evidence exists on the effects of mentoring on persistence or attrition in scientific disciplines. Information gathered from academe, industry, and the government should enable those institutions that have not yet implemented mentoring programs to learn from the successes of those who have taken such steps.
- Among suggestions for retaining women in science and engineering at various levels of the education/employment pipeline are the following:
  * providing financial assistance (for example, forgivable loan programs for students to complete graduate degree pro-

grams, grant-research programs to provide continuity between gaps in graduate support, and travel grants to attend conferences and to carry out research programs off campus) and equal access to scholarship resources;

* increasing the number and percentage of women with teaching and research assistantships;

* hiring junior and senior science majors to staff assistant positions; and

* increasing the number and percentage of women in science-related cooperative and intern programs (Connelly and Porter, 1978). These programs should be sensitive to sociocultural differences in order to attract and retain ethnic groups currently underrepresented in science and engineering. The Committee will undertake an evaluation to determine the access of women to such programs and their success in obtaining it.

• Using data already collected in national surveys of scientists and engineers (at all levels), the Committee on Women in Science and Engineering will study the degree attainment rates of women and men, particularly examining the transition from master's degree to doctorate, by field of study.

• Much anecdotal information has been put forward during the past five years about the negative influence of non-U.S. citizen faculty and graduate students on the recruitment and retention of female students in science and engineering, from the undergraduate through postdoctoral levels. With the increasing numbers of non-U.S. citizen faculty in some S&E disciplines, and as fewer U.S. citizens opt to pursue careers in these fields (and those who do make that career choice often prefer industrial employment), it is imperative that this issue be studied systematically. The Committee will develop a study of this phenomenon.

- The Committee, perhaps in conjunction with federal agencies for which OSEP administers postdoctoral research associateship programs, will study the present status of postdoctorals by sex, race/ethnicity, and academic discipline, including:

  * comparisons of fellowship versus research associateship appointments;
  * quality of postdoctoral experience;
  * size of stipends and the effects of stipends on beginning professional salaries, by field;
  * length of time in postdoctoral appointments; and
  * relation of postdoctoral experience to future permanent employment.

Such a study should form the basis of a reexamination of postdoctorals in science and engineering in the United States, updating information reported in *Postdoctoral Appointments and Disappointments* (1981) and reporting the effects of such training on gaining permanent employment.

- Using present data on the underemployment of women in scientific and engineering careers, the Committee will study this phenomenon. One goal would be to devise strategies to reduce the differential between women and men.

Monitoring Progress

- The Committee on Women in Science and Engineering will attempt to develop a model to project the magnitude of possible

110

growth in the number of women in the various S&E fields in the next decade, using currently available data. This model could then be used by the S&E community to determine future manpower estimates and appropriate interventions for providing an adequate supply of scientists and engineers for the year 2000 and beyond.

- The Committee will encourage all segments of the S&E education and policymaking communities to develop effective strategies for involving S&E faculty in both recruiting and retaining students, in order to make individual members of these communities aware of the importance of sociocultural and attitudinal factors on decisions by students initially interested in science or engineering studies to switch to nonscience/nonengineering fields.

- Because larger numbers of women are moving into industry, where a great shortage of personnel is predicted, the Committee will assist the industrial sector in taking steps to accommodate women and in making their careers "do-able" in the industrial context.

- Because the achievement and recognition of women scientists and engineers differs by country, the Committee on Women in Science and Engineering will examine government policies in the United States and abroad that have stimulated successful careers for women science and engineering, which in turn have increased the visibility of women scientists and engineers in universities, national laboratories, industry, and advisory posts.

Briefings

- The actions of media leaders, university administrators, employers, scientists and engineers, and the research community influence the participation of individuals in science and engineering. Meetings convened by the Committee on Women in Science and Engineering involve members of these groups, whose actions could create or influence change to overcome the underrepresentation of women in science and engineering. At these meetings sessions will deal with:

* changing U.S. demography;

* decreasing interest of American students (male and female) in careers in science and engineering;

* sociocultural and attitudinal factors that militate against the recruitment of women into science and engineering;

* difficulties that women encounter at the professional level; and

* efforts by all parties to increase the participation of women in science and engineering by eliminating gender-related training and employment inequities.

• The Committee will work with the media and the larger scientific community to develop strategies by which they can improve the public image of science and engineering and increase the interest of young people in these fields. One such project might be the development of a teenage science series patterned after the Nancy Drew mystery theme.

Dissemination of Information

• While recent research has led to a greater awareness of the factors affecting the recruitment and retention of women in science and engineering, this issue has many facets that warrant detailed study. Working with groups such as the Association of American College's Project on the Status and Education of Women, the Committee will develop a set of variables, such as the availability of mentoring and special scholarships and fellowship programs for women, to assess campus and corporate climate for women in science and engineering and to establish a database for these variables that may be accessible to individual and institutional researchers.

- Government subsidies or grants from private foundations for child care to undergraduate and graduate students and postdocs might also serve to recruit more women into scientific and engi- neering careers. The Committee will collect and disseminate infor-mation about successful programs to companies and academic institutions where child care considerations hamper employee recruitment, retention, performance, and morale.

- In some cases academe, industry, and government laboratories want to hire women professionals but claim to have difficulty identifying qualified women. The Committee will serve as a resource on highly qualified female scientists and engineers, relying on the Panelists File maintained in OSEP on individuals recommended to serve on the selection panels for the fellowship and associateship awards that OSEP administers. The biographical data include recent research topics.

- The doctoral surveys administered by NRC/OSEP are a valuable source of information on the characteristics and status of Ph.D.s. The Committee will use these data to analyze the status of women in science and engineering, by field and by employment sector, in order to make the policy community more aware of the career paths of women in science and engineering; their rates of promotion and tenuring in academe, industry, and government service in comparison to men's; and possible differences between fields and sectors of employment. The Committee will also attempt to disseminate field-specific data on application/acceptance ratios for fellowships and associateships, by sex and race, in addition to the current data collected on enrollment and retention rates.

In conclusion, the Committee stresses that while much is known about the participation of women in scientific and engineering careers and while we can derive great satisfaction over the improvements that have occurred in recent years, much remains to be accomplished. Deeper discussions of all issues delineated above should occur between practicing scientists and engineers, their professional societies, employers, the Office of Science and Technology Policy, the Congress, and the media, so as to meet head-on the challenges that face the United States in maintaining a competitive work force. The Committee views its role as that of a catalyst in bringing these diverse groups together in order to address the underparticipation of women in careers in the sciences and engineering.

BIBLIOGRAPHY

AFL-CIO. 1990. *Public Employees: Facts at a Glance.* Washington, D.C.: AFL-CIO Public Employee Department.

Ahern, Nancy C., and Elizabeth L. Scott. 1981. *Career Outcomes in a Matched Sample of Men and Women Ph.D.s: An Analytical Report.* Washington, D.C.: National Academy Press.

American Association of University Women. 1988. *The First Hundred Years.* Washington, D.C.: AAUW.

American Council on Education (ACE), Commission on Women in Higher Education. 1988. *Education the Majority: Women Challenge Tradition in Higher Education.* Washington, D.C.: ACE.

_____. 1991. *The ACE Fact Book on Women in Higher Education.* Washington, D.C.: ACE.

American Geological Institute (AGI). *Geotimes* (a monthly publication). Alexandria, Va.: AGI.

American Society for Engineering Education (ASEE). *Engineering Education* (a monthly publication). Washington, D.C.: ASEE.

Anderson, Mary R. 1990. Graduate reentry/career change program in industrial engineering. In *Women in Engineering Conference: A National Initiative* (conference proceedings). Jane Z. Daniels, ed. West Lafayette, Ind.: Purdue University.

Arnold, Karen. 1989. Retaining high-achieving women in science and engineering. In *Women in Science and Engineering: Changing Vision to Reality.* M. L. Matyas and S. M. Malcom, eds. Washington, D.C.: American Association for the Advancement of Science.

Association of American Colleges (AAC), Project on the Status and Education of Women (PSEW). *On Campus with Women* (a quarterly publication). Washington, D.C.: AAC.

Atkinson, Richard C. 1988. Bold steps are needed to educate the next generation of scientists. *The Chronicle of Higher Education*, March 2, p. B1.

Babco, Eleanor. 1985. *Salaries of Scientists, Engineers and Technicians.* Washington, D.C.: Scientific Manpower Commission (now the Commission on Professionals in Science and Technology).

Belsky, Jay. 1989. *Day Care Research: Implications for Employment.* Executive Summary of a presentation on "Marriage, Family, and Scientific Careers" at the annual meeting of the American

Association for the Advancement of Science, San Francisco, January 16.

Bogart, K. 1984. *Toward Equity: An Action Manual for Women in Academe.* Washington, D.C.: Association of American Colleges.

Bognanno, Mario F. 1987. Women in professions: Academic women. In *Working Women: Past, Present, Future.* Karen Shallcross Koziara, Michael H. Moskow, and Lucretia Dewey Tanner, eds. Washington, D.C.: The Bureau of National Affairs, Inc.

Brown, Robert W. 1990. Research apprenticeships for young undergraduates. In *Proceedings of the National Conference on Women in Mathematics and the Sciences.* Sandra Z. Keith and Philip Keith, eds. St. Cloud, Minn.: St. Cloud State University.

Burkheimer, Graham J., and Thomas P. Novak. 1981. *A Capsule Description of Young Adults Seven and One-Half Years After High School* (National Longitudinal Study Sponsored Reports Series (NCES 81-255). Research Triangle Park, N.C.: Center for Educational Research and Evaluation, Research Triangle Institute.

Campbell, Alan K., and Linda S. Dix, eds. 1990. *Recruitment, Retention, and Utilization of Federal Scientists and Engineers: A Report to the Carnegie Commission on Science, Technology, and Government.* Washington, D.C.: National Academy Press.

Campbell, Patricia B. 1991a. *Math, Science and Your Daughter: What Can Parents Do?* Groton, Mass.: Campbell-Kibler Associates.

____. 1991b. *Nothing Can Stop Us Now: Designing Effective Programs for Girls in Math, Science and Engineering.* Groton, Mass.: Campbell-Kibler Associates.

____. 1991c. *What Works and What Doesn't?* Groton, Mass.: Campbell-Kibler Associates.

____. 1991d. *Working Together, Making Changes: Working In and Out of School to Encourage Girls in Math and Science.* Groton, Mass.: Campbell-Kibler Associates.

Catalyst. 1986. *Career Development: A Partnership.* New York: Catalyst.

____. 1988. *Managing Work Force Diversity.* New York: Catalyst.

____. 1990. *Creative Mentoring.* New York: Catalyst.

____. 1991a. *Holding Managers Accountable for Promoting Women.* New York: Catalyst.

____. 1991b. *Honoring Corporate Initiatives to Promote Women's Leadership.* New York: Catalyst.

Cavanaugh, Margaret A. 1990. Strategies and programs for women in science and mathematics. In *Proceedings of the National Conference on Women in Mathematics and the Sciences.* Sandra Z. Keith and Philip Keith, eds. St. Cloud, Minn.: St. Cloud State University.

Chamberlain, Mariam K., ed. 1991. *Women in Academe: Progress and Prospects*. New York: Russell Sage Foundation.

Cheney, Lynne V. 1990. *Tyrannical Machines: A Report on Educational Practices Gone Wrong and Our Best Hopes for Setting Them Right*. Washington, D.C.: National Endowment for the Humanities.

Clemmitt, Marcia. 1990. Toughest federal science jobs elude women. *The Scientist* 4(20):8+, October 15.

Collins, Eileen C. 1988. Meeting the scientific and technical staffing requirements of the American economy. *Science and Public Policy* 15(5):335-342.

Committee on the Education and Employment of Women in Science and Engineering (CEEWISE). 1979. *Climbing the Academic Ladder: Doctoral Women Scientists in Academe*. Washington, D.C.: National Academy of Sciences.

_____. 1980. *Women Scientists in Industry and Government: How Much Progress in the 1970s?* Washington, D.C.: National Academy of Sciences.

_____. 1983. *Climbing the Ladder: An Update on the Status of Doctoral Women Scientists and Engineers*. Washington, D.C.: National Academy Press.

Committee on the Mathematical Sciences. 1990. *Renewing U.S. Mathematics: A Plan for the 1990s*. Washington, D.C.: National Academy Press.

Committee on Opportunities in the Hydrologic Sciences. 1991. *Opportunities in the Hydrologic Sciences*. Washington, D.C.: National Academy Press.

Committee on Research Opportunities in Biology. 1989. *Opportunities in Biology*. Washington, D.C.: National Academy Press.

Committee on Women's Employment and Related Social Issues (WERSI). 1986a. *Women's Work, Men's Work: Sex Segregation on the Job*. Washington, D.C.: National Academy Press.

____. 1986b. *Computer Chips and Paper Clips*. Washington, D.C.: National Academy Press.

____. 1988. *Sex Segregation in the Workplace: Trends, Explanations, and Remedies*. Washington, D.C.: National Academy Press.

Connelly, T., and A. L. Porter. 1978. *Women in Engineering: Policy Recommendations for Recruitment and Retention in Undergraduate Programs*. Atlanta: Georgia Institute of Technology.

Corning Professional Women's Forum. 1990. *Networking: A Guideline for Getting Started*. Corning, New York: Corning Incorporated.

Coyle, Susan L. 1986. *Summary Report 1985: Doctorate Recipients from United States Universities*. Washington, D.C.: National Academy Press, 1986. This edition of a report published annually by the

National Research Council's Office of Scientific and Engineering Personnel (OSEP) focused on women earning their doctorates in the sciences and engineering. Although not the focus of subsequent *Summary Reports,* data on such women in later cohorts is obtainable from OSEP's Doctorate Records File, which is funded jointly by the National Science Foundation, the National Institutes of Health, the U.S. Department of Education, the National Endowment for the Humanities, and the U.S. Department of Agriculture.

Daniels, Jane Z. 1987. A multifaceted approach to the recruitment of women into engineering. In *Contributions to the Fourth GASAT Conference, Volume One.* J. Z. Daniels and J. B. Kahle, eds. West Lafayette, Ind.: Purdue University.

Davis, Cinda-Sue. 1990. Networking at the University of Michigan Women in Science Program. In *Proceedings of the National Conference on Women in Mathematics and the Sciences.* Sandra Z. Keith and Philip Keith, eds. St. Cloud, Minn.: St. Cloud State University.

Dipboye, Robert L. 1987. Problems and progress of women in management. In *Working Women: Past, Present, Future.* Karen Shallcross Koziara, Michael H. Moskow, and Lucretia Dewey Tanner, eds. Washington, D.C.: The Bureau of National Affairs, Inc.

Dix, Linda S., ed. 1987. *WOMEN: Their Underrepresentation and Career Differentials in Science and Engineering* (Proceedings of a workshop). Washington, D.C.: National Academy Press.

Dybas, Cheryl. 1990. *American Science and Engineering Remain Productive Despite International Weaknesses* (NSF PR 90-7). Washington, D.C.: National Science Foundation.

Eagle, Eva, Robert A. Fitzgerald, Antoinette Gifford, and John Tuma. 1988. *National Longitudinal Study of 1972—A Descriptive Summary of 1972 High School Seniors: Fourteen Years Later* (CS 88-406). Washington, D.C.: U.S. Department of Education, Office of Education Research and Improvement.

Ehrhard, Julie Kuhn, and Bernice R. Sandler. 1987. *Looking for More Than a Few Good Women in Traditionally Male Fields.* Washington, D.C.: Association of American Colleges, Project on the Status and Education of Women.

Eisner, Robin. 1991. U.S. immigration law both helps and hinders foreign researchers. *The Scientist* 5(10):1+.

Engineering Manpower Commission. 1987. *Engineering and Technology Enrollments, Fall 1986.* Washington, D.C.: American Association of Engineering Societies.

Erickson, G. L., and L. G. Erickson. 1984. Females and science achievement: Evidence, explanations, and implications. *Science Education* **68**:63-89.

Falk, Charles E. 1990. Quantitative inputs to federal technical personnel management. In *Recruitment, Retention, and Utilization of Federal Scientists and Engineers: A Report to the Carnegie Commission on Science, Technology, and Government.* Alan K. Campbell and Linda S. Dix, eds. Washington, D.C.: National Academy Press.

Federal Coordinating Council for Science, Engineering, and Technology (FCCSET), Committee on Education and Human Resources. 1991. *By the Year 2000: First in the World.* Washington, D.C.: Executive Office of the President, Office of Science and Technology Policy.

Finholt, James E. 1990. Carleton chemistry: A success story for women. In *Proceedings of the National Conference on Women Mathematics and the Sciences.* Sandra Z. Keith and Philip Keith, eds. St. Cloud, Minn.: St. Cloud State University.

Galinsky, Ellen. 1988. Child Care and Productivity. Paper prepared for the Child Care Action Campaign conference, "Child Care: The Bottom Line," New York, NY.

____, and Peter Stein. 1989. *Balancing Careers and Families: Research Findings and Institutional Responses.* Presentation in the session "Marriage, Family, and Scientific Careers" at the annual meeting of the American Association for the Advancement of Science, San Francisco, January 16.

General Mills Survey. 1981. *Families at Work: The General Mills American Report.* Minneapolis: General Mills.

Gilbert, L. A. 1985. *Men in Dual-Career Families,* Hillsdale, N.J.: Lawrence Erlbaum Associates.

Gilford, Dorothy M., and Ellen Tenenbaum, eds. 1990. *Precollege Science and Mathematics Teachers: Monitoring Supply, Demand, and Quality.* Washington, D.C.: National Academy Press.

Gordon, Henry A. 1990. *Who Majors in Science? College Graduates in Science, Engineering, or Mathematics from the High School Class of 1980* (NCES 90-658). Washington, D.C.: U.S. Government Printing Office.

Government-University-Industry Research Roundtable (GUIRR). 1987. *Nurturing Science and Engineering Talent: A Discussion Paper.* Washington, D.C.: GUIRR

Hall, Roberta M. 1982. *The Classroom Climate: A Chilly One for Women?* Washington, D.C.: Association of American Colleges' Project on the Status and Education of Women.

____, and Bernice R. Sandler. 1983. *Academic Mentoring for Women Students and Faculty: A New Look at an Old Way to Get Ahead.*

Washington, D.C.: Association of American Colleges' Project on the Status and Education of Women.

Hargens, Lowell L., Jr. 1971. *The Social Context of Scientific Research,* Ph.D. dissertation, University of Wisconsin.

Harvard University. 1991. Outnumbered and underconnected. *Harvard Magazine* (May-June):74-76.

Hornig, Lilli S. 1987. Women graduate students: A literature review and synthesis. In *WOMEN: Their Underrepresentation and Career Differentials in Science and Engineering.* Linda S. Dix, ed. Washington, D.C.: National Academy Press.

How much science is enough? (editorial) *The Washington Post.* April 24, 1991.

Johnson, Robert C. 1990. Black female participation in quantitative domains. In *Proceedings of the National Conference on Women in Mathematics and the Sciences.* Sandra Z. Keith and Philip Keith, eds. St. Cloud, Minn.: St. Cloud State University.

Jones, Marjorie A. 1990. Numbers of women in graduate school in the sciences—A case for concern? In Sandra Z. Keith and Philip Keith, eds. *Proceedings of the National Conference on Women in Mathematics and the Sciences.* St. Cloud, Minn.: St. Cloud State University.

Kahle, Jane Butler, and Marsha Lakes Matyas. 1987. Equitable science and mathematics education: A discrepancy model. In *WOMEN: Their Underrepresentation and Career Differentials in Science and Engineering.* Linda S. Dix, ed. Washington, D.C.: National Academy Press.

Kay, Nina. 1990. Introductory remarks. Symposium on Women in Science, Engineering, and Technology, Houston, Tex., October 14, 1990.

Keith, Sandra Z., and Philip Keith, eds. 1990. *Proceedings of the National Conference on Women in Mathematics and the Sciences.* St. Cloud, Minn.: St. Cloud State University.

Kennedy, Dale. 1990. The Douglass Project for Rutgers women in math and science. In *Proceedings of the National Conference on Women in Mathematics and the Sciences.* Sandra Z. Keith and Philip Keith, eds. St. Cloud, Minn.: St. Cloud State University.

Knapp, Edward A. 1983. Comments at the Conference on New Directions for Research on Women. Washington, D.C.: National Academy of Sciences, September 9.

Koshland, Daniel E., Jr. 1988. *Science* 239(4787):1473, March 25.

Koziara, Karen Shallcross. 1987. Women and work: The evolving policy.In *Working Women: Past, Present, Future.* Karen ShallcrossKoziara, Michael H. Moskow, and Lucretia Dewey

Tanner, eds. Washington, D.C.: The Bureau of National Affairs, Inc.

_____, Michael H. Moskow, and Lucretia Dewey Tanner, eds. 1987. *Working Women: Past, Present, Future.* Washington, D.C.: The Bureau of National Affairs, Inc.

Latham, G. P., and K. N. Wesley. 1981. *Increasing Productivity Through Performance Appraisal.* Reading, Mass.: Addison-Wesley.

LeBold, William K. 1987. Women in engineering and science: An undergraduate research perspective. In *WOMEN: Their Underrepresentation and Career Differentials in Science and Engineering.* Linda S. Dix, ed. Washington, D.C.: National Academy Press.

Lee, Carolyn S., ed. 1991. *Mathematics Education Programs That Work* (PIP91-835). Washington, D.C.: U.S. Department of Education, Office of Educational Research and Improvement.

MacCorquodale, Patricia. 1984. *Self-Image, Science, and Math: Does the Image of the "Scientist" Keep Girls and Minorities from Pursuing Science and Math?"* (NIE-G-79-0111). Paper presented at the annual meeting of the American Sociological Association, San Antonio, Texas.

Malcom, Shirley M. 1983a. *Equity and Excellence: Compatible Goals.* Washington, D.C.: American Association for the Advancement of Science.

_____. 1983b. Untitled keynote address. Conference on New Directions for Research on Women in Science and Engineering. Washington, D.C.: National Academy of Sciences, September 9.

Mappen, Ellen F. 1990. The Douglass Project for Rutgers women in math, science, and engineering: A comprehensive program to encourage women's persistence in these fields. In *Women in Engineering Conference: A National Initiative* (conference proceedings). Jane Z. Daniels, ed. West Lafayette, Ind.: Purdue University.

Mathematical Sciences Education Board (MSEB) and the Board on Mathematical Sciences (BMS). 1989. *Everybody Counts: A Report to the Nation on the Future of Mathematics Education.* Washington, D.C.: National Academy Press.

Matyas, Marsha Lakes. 1987. *Partial List of Precollege Mathematics and Science Programs for Minority and/or Female Students.* Washington, D.C.: American Association for the Advancement of Science, Office of Opportunities in Science.

_____, and Shirley M. Malcom, eds. 1991. *Investing in Human Potential: Science and Engineering at the Crossroads.* Washington, D.C.: American Association for the Advancement of Science (AAAS).

Mauersberger, Konrad. 1990. Performance of female and male students in a three-quarter general physics course. In *Proceedings of the National Conference on Women in Mathematics and the Sciences.* St. Cloud, Minn.: Sandra Z. Keith and Philip Keith, eds. St. Cloud State University.

McKee, Marie. 1991. *Increasing the Participation of Women in Engineering Careers: A Corporate Perspective.* Keynote address at the 2nd annual Women in Engineering Conference: A National Initiative, Washington, D.C., June 4.

McLaren, Christine E. 1990. Undergraduate research experiences in mathematics. In *Proceedings of the National Conference on Women in Mathematics and the Sciences.* Sandra Z. Keith and Philip Keith, eds. St. Cloud, Minn.: St. Cloud State University.

Megaw, W. J. 1991. *Gender Distribution in the World's Physics Departments.* Paper prepared for the meeting, Gender and Science and Technology 6, Melbourne, Australia, July 14-18.

Merton, Robert K. 1973. The normative structure of science. *The Sociology of Science.* Chicago: University of Chicago Press, pp. 267-278.

Metzler-Brennan, Elizabeth, Robin Lewis, and Meg Gerrard. 1985. Childhood antecedents of women's masculinity, femininity, and career role choices. *Psychology of Women Quarterly* 9:371-382.

Moran, Mary. 1986. *Student Financial Aid and Women: Equity Dilemma.* Washington, D.C.: Association for the Study of Higher Education.

Mulnix, Amy. 1990. Change the woman or change the graduate school? In *Proceedings of the National Conference on Women in Mathematics and the Sciences.* Sandra Z. Keith and Philip Keith, eds. St. Cloud, Minn.: St. Cloud State University.

Nagy, Theresa A., and Ellen Cunningham. 1990. The role model as the embodiment of a worked example. In *Proceedings of the National Conference on Women in Mathematics and the Sciences.* Sandra Z. Keith and Philip Keith, eds. St. Cloud, Minn.: St. Cloud State University.

National Academy of Sciences (NAS), Committee on the Conduct of Science. 1989. *On Being a Scientist.* Washington, D.C.: National Academy Press.

National Research Council. 1981. *Postdoctoral Appointments and Disappointments.* Washington, D.C.: National Academy Press.

___. 1989. *High School Biology Today and Tomorrow.* Washington, D.C.: National Academy Press.

National Science Board. 1989. Science and Engineering Indicators:——1989 (NSB89-1). Washington, D.C.: U.S. Government Printing Office.

National Science Foundation (NSF). 1987. *Foreign Citizens in U.S.*

Science and Engineering: History, Status and Outlook. Washington, D.C.: U.S. Government Printing Office.

———. 1988. *Women and Minorities in Science and Engineering* (NSF 88-301). Washington, D.C.: National Science Foundation.

———. 1989. *Federal Scientists and Engineers: 1988.* Washington, D.C.: U.S. Government Printing Office.

———. 1990a. *Report of the Workshop on the Dissemination and Transfer of Innovation in Science, Mathematics, and Engineering Education* (NSF 91-21). Washington, D.C.: Directorate for Education and Human Resources.

———. 1990b. *Women and Minorities in Science and Engineering* (NSF 90-301). Washington, D.C.: National Science Foundation.

National Science Resources Center, Smithsonian Institution, and National Academy of Sciences. 1988. *Science for Children, Resources for Teachers.* Washington, D.C.: National Academy Press.

Nieva, Veronica F. 1985. Work and family linkages. In *Women and Work: An Annual Review,* Volume 1. Laurie Larwood, Ann H. Stromberg, and Barbara A. Gutek, eds. Beverly Hills, Calif.: Sage Publications.

NSF, universities plan for women, minorities. *Chronicle of Higher Education.* September 5, 1990, p. A2.

Oberlin College. 1986a. Executive Summary, *The Future of Science at Liberal Arts Colleges* (Report of a conference held June 9-10, 1985). Oberlin, Ohio: Office of the Provost.

———. 1986b. *Maintaining America's Scientific Productivity: The Necessity of the Liberal Arts Colleges.* Oberlin, Ohio: Office of the Provost.

Palguta, John M. 1990. Meeting federal work force needs with regard to scientists and engineers: The role of the U.S. Office of Personnel Management. In *Recruitment, Retention, and Utilization of Federal Scientists and Engineers: A Report to the Carnegie Commission on Science, Technology, and Government.* Alan K. Campbell, and Linda S. Dix, eds. Washington, D.C.: National Academy Press.

Pine, Patricia. 1987. Before they are undergraduates. *Mosaic* 18(1):34-44.

Pleck, J., G. Staines, & L. Long. 1978. *Work and Family Life: First Reports on Work-Family Interferences and Workers' Formal Child Care Arrangement from the 1977 Quality of Employment Survey.* Ann Arbor, Mich.: University of Michigan.

Press, Frank. 1988. The dilemma of the Golden Age. Address to the members, National Academy of Sciences, Washington, D.C., April 26.

Rawls, Rebecca L. 1991. Minorities in science. *Chemical & Engineering News* 69(15):20-35.

Rayman, Paula, and Heather Burbage. 1989. *Professional Families:*

Falling Behind While Getting Ahead. Presentation in the session "Marriage, Family, and Scientific Careers" at the annual meeting of the American Association for the Advancement of Science, San Francisco, January 16.

Reilly, William K. 1990. Toward the year 2000: EPA's study on cultural diversity in its workforce. A Memorandum to [EPA] headquarters managers and supervisors. May 7.

Reskin, Barbara F. 1976. Sex differences in status attainment in science: The case of the postdoctoral fellowship. *American Sociological Review* 41:597-612.

___, and Heidi I. Hartmann, eds. 1986. *Women's Work, Men's Work: Sex Segregation on the Job.* Washington, D.C.: National Academy Press.

Roby, P. 1973. Institutional barriers to women students in higher education. In *Academic Women on the Move.* Alice S. Rossi and Ann Calderwood, eds. New York: Russell Sage Foundation.

Rosenfeld, Rachel A., and James C. Hearn. 1982. Sex difference in the significance of economic resources for choosing and attending college. In *The Undergraduate Woman: Issues in Education Equity.* P. Perena, ed. Lexington, Mass.: Lexington, Books.

Sandler, Bernice R. 1986. *The Campus Climate Revisited: Chilly for Women Faculty, Administrators, and Graduate Students.* Washington, D.C.: Association of American Colleges' Project on the Status and Education of Women.

Schellhardt, Timothy D., and Carol Hymowitz. 1989. For Roderick, the future lies in factories. *The Wall Street Journal,* May 22, p. B8.

Schonberger, Ann K. 1990. College women's persistence in engineering and physical science. In *Proceedings of the National Conference on Women in Mathematics and the Sciences.* Sandra Z. Keith and Philip Keith, eds. St. Cloud, Minn.: St. Cloud State University.

Segal, Edward. 1991. Homework. *Government Executive* 23(5):46-49.

Selvin, Paul. 1991. Does the Harrison case reveal sexism in math? *Science* 252:1781-1783.

Shell, K. D., W. K. LeBold, K. W. Linden, and C. M. Jagacinski. 1985. Retention research in engineering education. In *Proceedings of the 1985 Frontiers in Education Conference.* Arlington, Texas: American Society for Engineering Education.

Silvertsen, Mary Lewis, ed. 1990. *Science Education Programs That Work* (PIP 90-846). Washington, D.C.: U.S. Department of Education, Office of Educational Research and Improvement.

Simon, Lisa. 1991. What are the goals and priorities of the average scientist? *The Scientist* 5(14):20.

Smith, Page. 1990. *Killing the Spirit: Higher Education in America.* New York: Viking Press.

Speizer, J. J. 1981. Role models, mentors, and sponsors: The elusive concepts. *Signs* 6:692-712.

Task Force on Women, Minorities, and the Handicapped in Science and Technology (Task Force). 1988. *Changing America: The New Face of Science and Engineering* (Interim Report). Washington, D.C.: The Task Force.

____. 1989. *Changing America: The New Face of Science and Technology* (Final Report). Washington, D.C.: The Task Force.

Thurgood, Delores H., and Joanne M. Weinman. 1990. *Summary Report 1989: Doctorate Recipients from United States Universities.* Washington, D.C.: National Academy Press. (Biennial data from the Survey of Earned Doctorates is collected and maintained within the National Research Council's Office of Scientific and Engineering Personnel.)

Touchton, Judith G., and Lynne Davis, comp. 1991. *1990-91 Fact Book on Higher Education.* New York: American Council on Education and Macmillan Publishing Company.

Transportation Research Board. 1985. *Transportation Education and Training: Meeting the Challenge.* Proceedings of the Conference on Surface Transportation Education and Training, Williamsburg, Va., October 28-31, 1984. Washington, D.C.: Transportation Research Board.

Tuckman, Howard, Susan Coyle, and Yupin Bae. 1990. *On Time to the Doctorate: A Study of the Increased Time to Complete Doctorates in Science and Engineering.* Washington, D.C.: National Academy Press.

U.S. Congress, Office of Technology Assessment (OTA). 1985. *Demographic Trends and the Scientific and Engineering Work Force—A Technical Memorandum.* Washington, D.C.: U.S. Government Printing Office.

____. 1988. *Educating Scientists and Engineers: Grade School to Grad School* (OTA-SET-377). Washington, D.C.: U.S. Government Printing Office.

____. 1990. *Proposal Pressure in the 1980s: An Indicator of Stress on the Federal Research System.* A staff paper. Washington, D.C.: OTA.

U.S. Department of Education, National Center for Education Statistics (NCES). 1970+. *Earned Degrees Conferred* series.

U.S. Department of Energy (DOE), Office of University and Science Education Programs. 1991. *Review of Laboratory Programs for Women.* Summary of a meeting hosted by Argonne National Laboratory and the University of Chicago, November 16, 1990.

U.S. Department of Labor, Bureau of Labor Statistics. 1990. *Outlook 2000* (Bulletin 2352). Washington, D.C.: U.S. Government Printing Office.

U.S. Department of Labor, Women's Bureau. 1988. *Work and Family Resource Kit.* Washington, D.C.: The Bureau.

U.S. Office of Personnel Management (OPM). 1988. *Helping Federal Employees Balance Work and Family Life.* Washington, D.C.: OPM.

___. 1990. *Flexiplace Focus: The Federal Flexiplace Newsletter.* April. Washington, D.C.: OPM.

University of California, Berkeley. 1989. *Report of the Coordinating Committee on the Status of Women.* The University: Office of the President.

Vetter, Betty M. 1987. Women's progress. *Mosaic* 18(1):2-9.

___. 1989a. Replacing engineering faculty in the 1990s. Engineering Education 79(5):540-546.

___. 1989b. *Women in Science: Progress and Problems.* Occasional Paper 89-1. Washington, D.C.: Commission on Professionals in Science and Technology.

___. 1991. *Professional Women and Minorities: A Manpower Data Resource Service.* Washington, D.C.: Commission on Professionals in Science and Technology.

Welch, John F., Jr. 1991. Restoring upward mobility. *Issues in Science and Technology* VII(3):38-40.

Zuckerman, Harriet. 1987. Persistence and change in the careers of men and women scientists and engineers: A review of current research. In Linda S. Dix, ed. *WOMEN: Their Underrepresentation and Career Differentials in Science and Engineering.* Washington, D.C.: National Academy Press.

Zumeta, William M. 1985. *Extending the Educational Ladder: The Changing Quality and Value of Postdoctoral Study.* Lexington, Mass.: Lexington Books.

RELATED TABLES

A: Top 25 Science and Engineering Degree-Granting
      Institutions, 1980-1990   128
B: NSF Graduate Fellowship Program Applications
      and Awards, by Sex, 1985-1991   130
C: NSF Graduate Fellowship Program Applications by
      Women, by Ethnic Group, 1989-1991   134
D: NSF Graduate Fellowship Program Awards to Women, by
      Ethnic Group, 1989-1991   136
E: NSF Minority Graduate Fellowship Program Applications
      by Women, by Ethnic Group, 1989-1991   138
F: NSF Minority Graduate Fellowship Program Awards to
      Women, by Ethnic Group, 1989-1991   140

TABLE A: Top 25 Science and Engineering Doctorate-Granting Institutions, 1980-1990

Institution	1980	1981	1982	1983	1984	1985	1986	1987	1988	1989	1990	Total 1980-1990
TOTAL, MALE												
Calif, U-Berkeley	439	384	432	411	422	421	427	408	457	492	471	4764
Ill, U, Urbana-Champ	355	354	312	327	325	354	305	362	367	377	408	3846
Mass Inst Technology	326	325	343	347	337	350	357	356	404	390	393	3928
Cornell Univ/NY	276	270	259	281	264	262	284	285	263	284	347	3075
Wisconsin, U-Madison	321	316	349	315	317	353	326	369	362	370	342	3740
Stanford Univ/CA	278	318	284	259	305	269	334	327	348	333	323	3378
Purdue University/IN	260	285	249	251	261	249	252	230	253	282	312	2884
Minnesota, U-Minneapolis	223	231	232	208	238	265	289	210	259	269	300	2724
Michigan, Univ of	239	236	241	268	259	287	256	249	268	269	290	2861
Texas, U-Austin	168	185	185	175	174	193	238	257	250	245	283	2353
Calif, U-Los Angeles	228	261	232	237	218	216	198	200	256	231	271	2548
Ohio State Univ	238	195	239	223	205	235	231	250	228	278	264	2586
Texas A&M University	182	169	158	168	196	192	186	215	212	246	242	2166
Michigan State Univ	233	230	249	237	197	194	180	199	188	220	220	2347
Penn State Univ	181	182	193	204	196	191	182	191	205	228	219	2172
Florida, Univ of	137	134	112	154	166	176	162	173	179	184	213	1790
Harvard Univ/MA	195	180	179	195	179	155	178	172	181	164	209	1987
Maryland, Univ of	135	120	129	129	139	138	147	150	149	165	206	1607
Washington, U of	174	185	194	192	174	171	191	202	198	205	201	2087
NC State U-Raleigh	98	106	140	141	143	135	155	147	162	167	199	1593
Arizona, Univ of	145	133	154	151	168	143	139	169	171	184	184	1741
Calif, U-Davis	184	200	151	229	184	155	170	162	182	192	178	1987
Iowa State Univ	168	184	176	157	161	166	168	207	191	181	170	1929
Northwestern Univ/IL	125	139	151	130	135	161	147	149	159	184	168	1648
VA Poly Inst&State U	101	117	111	133	132	140	146	151	150	207	166	1554

128

Institution	1980	1981	1982	1983	1984	1985	1986	1987	1988	1989	1990	Total 1980-1990
TOTAL, FEMALE												
Calif, U-Berkeley	101	99	110	108	95	128	136	126	128	166	136	1333
Wisconsin, U-Madison	79	73	93	83	82	105	87	84	117	115	118	1036
Calif, U-Los Angeles	86	69	78	74	78	80	84	88	107	103	111	958
Cornell Univ/NY	54	63	69	74	87	75	87	82	114	111	108	924
Ill, U, Urbana-Champ	39	57	46	55	56	88	78	72	77	91	106	765
Minnesota, U-Minneapol	68	85	59	67	74	71	92	100	78	91	103	888
Ohio State Univ	62	79	66	70	64	88	66	72	79	93	102	841
Maryland, Univ of	40	52	73	63	70	72	66	70	56	79	95	736
Michigan, Univ of	53	79	82	108	89	84	92	95	76	67	92	917
Stanford Univ/CA	49	40	55	49	59	65	59	78	63	79	88	684
Mass Inst Technology	46	59	48	65	52	71	85	80	85	79	85	755
Pennsylvania, U of	46	58	75	72	77	50	58	81	73	78	85	753
Texas, U-Austin	51	43	49	53	53	62	60	73	76	84	84	688
CUNY-Grad Sch&U Ctr	48	58	54	61	62	48	65	73	79	78	82	708
Washington, U of	54	46	52	57	63	50	58	60	81	67	82	670
NC, U of-Chapel Hill	66	51	42	63	49	56	42	61	65	53	81	629
Columbia Univ/NY	85	68	62	63	78	73	81	65	72	86	79	812
Purdue University/IN	37	41	39	54	47	59	68	70	49	66	71	601
Rutgers Univ/NJ	35	61	35	54	46	44	47	58	53	64	68	565
Northwestern Univ/IL	53	45	42	49	53	57	60	62	61	75	66	623
Michigan State Univ	48	49	56	62	53	46	62	59	80	67	66	648
New York University	83	74	82	66	75	70	68	74	70	55	65	782
Penn State Univ	34	51	47	57	47	43	55	60	55	61	63	573
Texas A&M University	29	26	22	34	31	29	42	42	41	64	63	423
Mass, U of-Amherst	42	36	37	43	48	46	48	45	48	52	61	506

SOURCE: National Science Foundation, unpublished data.

129

TABLE B: NSF Graduate Fellowship Program Applications and Awards, by

| Discipline | 1985 | | 1986 | | 1987 |
	Men	Women	Men	Women	Men
Total Applicants					
N	2776	1614	3063	1816	2924
%	63.2	36.8	62.8	37.2	61.9
Biochemistry*	246	167	243	186	262
	59.6	40.4	56.6	43.4	58.1
Biology	298	274	304	297	285
	52.1	42.9	50.6	49.4	52.1
Chemistry	219	118	269	110	230
	65.0	35.0	71.0	29.0	64.8
Earth Sciences	151	88	134	90	110
	63.2	36.8	59.8	40.2	62.5
Applied Math/Statistics	80	39	68	32	73
	67.2	32.8	68.0	32.0	52.9
Mathematics	105	43	116	48	134
	70.9	29.1	70.7	29.3	70.5
Physics and Astronomy	309	44	298	54	250
	87.5	12.5	84.7	15.3	81.2
Behavioral Sciences**	397	436	467	508	468
	47.7	52.3	47.9	52.1	49.2
Biomedical Sciences	154	208	171	214	166
	42.5	57.5	44.4	55.6	43.5
Computer Science	182	54	236	64	221
	77.1	22.9	78.7	21.3	76.5
Engineering	635	143	757	213	725
	81.6	18.4	78.0	22.0	77.1

Sex, 1985-1991

Women	1988		1989		1990		1991	
	Men	Women	Men	Women	Men	Women	Men	Women
1802	3145	2006	3238	2129	3629	2680	4145	3201
38.1	61.1	38.9	60.3	39.7	57.5	42.5	56.4	43.6
189	257	230	261	254	268	249	276	256
41.9	52.8	47.2	50.7	49.3	51.8	48.2	51.9	48.1
262	299	303	263	335	288	350	369	432
47.9	49.7	50.3	44.0	56.0	45.1	54.9	46.1	53.9
125	246	156	220	149	237	149	272	174
35.2	61.2	38.8	59.6	40.4	61.4	38.6	61.0	39.0
66	117	56	125	57	117	83	116	124
37.5	67.6	32.4	68.7	31.3	58.5	41.5	48.3	51.7
65	71	55	90	47	82	69	100	87
47.1	56.3	43.7	65.7	34.3	54.3	45.7	53.5	46.5
56	134	45	127	52	157	67	132	90
29.5	74.9	25.1	70.9	29.1	70.1	29.9	59.5	40.5
58	324	56	367	78	396	86	404	99
18.8	85.3	14.7	82.5	17.5	82.2	17.8	80.3	19.7
483	483	565	513	620	628	690	627	780
50.8	46.1	53.9	45.3	54.7	47.6	52.4	44.6	55.4
216	196	261	164	212	188	261	195	311
56.5	42.9	57.1	43.6	56.4	41.9	58.1	38.5	61.5
68	223	48	244	67	254	88	282	67
23.5	82.3	17.7	78.5	21.5	74.3	25.7	80.8	19.2
215	795	231	864	258	1014	588	1280	692
22.9	77.5	22.5	77.0	23.0	63.3	36.7	64.9	35.1

TABLE B (continued)

Discipline	1985		1986		1987
	Men	Women	Men	Women	Men
Total Awards					
N	362	178	332	173	327
%	67.0	33.0	66.0	34.0	65.0
Biochemistry*	32	16	21	22	27
	66.7	33.3	48.8	51.2	54.0
Biology	32	40	28	29	29
	44.4	55.6	49.1	50.9	54.7
Chemistry	32	9	33	8	27
	78.0	22.0	80.5	19.5	73.0
Earth Sciences	20	9	17	9	11
	69.0	31.0	65.4	34.6	64.7
Applied Math/Statistics	14	1	8	1	10
	93.3	6.7	88.9	11.1	67.7
Mathematics	19	1	16	2	21
	95.0	5.0	88.9	11.1	95.5
Physics and Astronomy	39	6	34	6	32
	86.7	13.3	85.0	15.0	86.5
Behavioral Sciences**	50	50	50	47	55
	50/0	50.0	51.5	48.5	55.6
Biomedical Sciences	15	28	16	18	14
	42.5	57.5	44.4	55.6	43.5
Computer Science	27	3	27	6	21
	90.0	10.0	81.8	18.2	61.8
Engineering	82	15	82	25	80
	84.5	15.5	76.6	23.4	73.4

* Includes biochemistry, biophysics, and molecular biology.
**Prior to 1991, this field included psychology, economics, and sociology.
ogy, and linguistics; (2) economics, urban planning, and history of science;
——did not occur until 1991, a single category is used here.
SOURCE: Office of Scientific and Engineering Personnel.

	1988		1989		1990		1991	
Women	Men	Women	Men	Women	Men	Women	Men	Women
178	440	245	489	271	494	357	556	394
35.0	64.0	36.0	64.3	35.7	58.0	42.0	58.5	41.5
23	31	35	36	36	38	25	31	31
46.0	47.0	53.0	50.0	50.0	60.3	39.7	50.0	50.0
24	41	32	35	44	28	47	40	53
45.3	56.2	43.8	44.3	55.7	37.3	62.7	43.0	57.0
10	31	23	35	19	29	20	41	16
27.0	57.4	42.6	64.8	35.2	59.2	40.8	71.9	28.1
6	16	8	15	8	17	9	13	16
35.3	66.7	33.3	65.2	34.8	65.4	34.6	44.8	55.2
5	13	4	16	6	18	4	18	4
33.3	76.5	23.5	72.7	27.3	81.8	18.2	81.8	18.2
1	25	3	26	4	28	6	22	10
4.5	89.3	10.7	86.7	13.3	82.4	17.6	68.8	31.2
5	54	5	62	13	59	9	57	13
13.5	91.5	8.5	82.7	17.3	86.8	13.2	81.4	18.6
44	64	68	79	65	62	46	92	89
44.4	48.5	51.2	54.9	45.1	57.4	42.6	50.8	49.2
19	16	31	13	28	19	25	23	30
56.5	42.9	57.1	31.7	68.3	43.2	56.8	43.4	56.6
13	34	6	40	10	34	11	40	5
38.2	85.0	15.0	80.0	20.0	75.6	24.4	88.9	11.1
29	115	30	132	38	106	76	179	127
26.6	79.3	20.7	77.6	22.4	58.2	41.8	58.5	41.5

Because the disaggregation of behavioral sciences——into (1) anthropology, sociol-
(3) political science, international relations, and geography; and (4) psychology

133

TABLE C: NSF Graduate Fellowship Program Applications by Women, by Ethnic Group, 1989-1991

Discipline	American Indian			Asian			Black			White			Pacific Islander		
	1989	1990	1991	1989	1990	1991	1989	1990	1991	1989	1990	1991	1989	1990	1991
Grand Total, Women	2129	2680	3201	2129	2680	3201	2129	2680	3201	2129	2680	3201	2129	2680	3201
Ethnic Subtotal	12 / .6	19 / .7	25 / .8	136 / 6.4	161 / 6.0	191 / 6.0	198 / 9.3	235 / 8.8	356 / 11.1	1608 / 75.5	2082 / 77.7	2373 / 74.1	18 / .8	19 / .7	17 / .5
Biochemistry	1	1	1	24 / 1.1	22 / .8	27 / .8	16 / .8	17 / .6	21 / .7	194 / 9.1	196 / 7.3	192 / 6.0	3 / .1	1	1
Biology	2 / .1	1	4 / .1	8 / .4	12 / .4	12 / .4	20 / .9	14 / .5	23 / .7	283 / 13.3	298 / 11.1	365 / 11.4	4 / .2	2 / .1	1
Chemistry		1	1	7 / .3	12 / .4	16 / .5	13 / .6	12 / .4	12 / .4	123 / 5.8	113 / 4.2	130 / 4.1	3 / .1	1	1
Earth Sciences		1	1	1	1	3 / .1	5 / .2	1	2 / .1	48 / 2.3	73 / 2.7	108 / 3.4			
Applied Math/Statistics		1	1	5 / .2	6 / .2	6 / .2	6 / .3	8 / .3	17 / .5	32 / 1.5	53 / 2.0	58 / 1.8			1
Mathematics			1		1	4 / .1	7 / .3	3 / .1	6 / .2	44 / 2.1	57 / 2.1	73 / 2.3		2 / .1	2 / .1
Physics and Astronomy				9 / .4	8 / .3	8 / .2	1	3 / .1	4 / .1	63 / 3.0	73 / 2.7	84 / 2.6		1	
Behavioral Sciences	7 / .3	7 / .3	6 / .2	30 / 1.4	23 / .9	39 / 1.2	71 / 3.3	88 / 3.3	121 / 3.8	447 / 21.0	518 / 19.3	613 / 19.2	4 / .2	3 / .1	5 / .2
Biomedical Sciences	1	2 / .1	1	10 / .5	8 / .3	13 / .4	29 / 1.4	28 / 1.0	39 / 1.2	149 / 7.0	201 / 7.5	226 / 7.1	1	1	2 / .1
Computer Sciences			2 / .1	8 / .4	6 / .2	3 / .1	9 / .4	17 / .6	14 / .4	44 / 2.1	61 / 2.3	41 / 1.3	2 / .1	1	1
Engineering	1	4 / .1	4 / .1	34 / 1.6	62 / 2.3	60 / 1.9	21 / 1.0	44 / 1.6	97 / 3.0	181 / 8.5	439 / 16.4	483 / 15.1	1	7 / .3	4 / .1

134

TABLE C (continued)

Discipline	Mexican American			Puerto Rican			Cuban			Other Hispanic			Unspecified		
	1989	1990	1991	1989	1990	1991	1989	1990	1991	1989	1990	1991	1989	1990	1991
Grand Total	2129	2680	3201	2129	2680	3201	2129	2680	3201	2129	2680	3201	2129	2680	3201
Ethnic Subtotal	41 1.9	53 2.0	66 2.1	50 2.3	37 1.4	64 2.0	1	8 .3	21 .7	45 2.1	41 1.5	64 2.0	20 .9	25 .9	24 .7
Biochemistry	3 .1	4 .1	4 .1	7 .3	4 .1	4 .1			1	3 .1	3 .1	2 .1	3 .1	1	3 .1
Biology	6 .3	10 .4	8 .2	3 .1	4 .1	3 .1			4 .1	6 .3	5 .2	9 .3	3 .1	4 .1	3 .1
Chemistry	1		4 .1	1	6 .2	6 .2		1			2 .1	2 .1	1	1	1
Earth Sciences		1	2 .1	2 .1	3 .1	2 .1			1	2 .1	1	1	1	1	
Applied Math/Statistics	1	1	1			1							1		
Mathematics			2 .1			2 .1				2 .1	2 .1	2 .1		2 .1	
Physics and Astronomy	2 .1	1	1			1	1			1			2 .1		
Behavioral Sciences	13 .6	22 .8	27 .8	21 1.0	7 .3	21 .7		3 .1	7 .2	22 1.0	9 .3	24 .7	5 .2	10 .4	6 .2
Biomedical Sciences	4 .2	7 .3	6 .2	11 .5	7 .3	13 .4		1		5 .2	6 .2	9 .3	2 .1		2 .1
Computer Sciences	1		1	2 .1		2 .1			1	1	3 .1	1	1		1
Engineering	10 .5	7 .3	10 .3	3 .1	6 .2	9 .3		3 .1	6 .2	5 .2	10 .4	13 .4	2 .1	6 .2	6 .2

* Includes biochemistry, biophysics, and molecular biology.
** Prior to 1991, this field included psychology, economics, and sociology. Because the disaggregation of behavioral sciences——into (1) anthropology, sociology, and linguistics; (2) economics, urban planning, and history of science; (3) political science, international relations, and geography; and (4) psychology——did not occur until 1991, a single category is used here.
SOURCE: Office of Scientific and Engineering Personnel.

TABLE D: NSF Graduate Fellowship Program Awards to Women, by Ethnic Group, 1989-1991

Discipline	American Indian 1989	American Indian 1990	American Indian 1991	Asian 1989	Asian 1990	Asian 1991	Black 1989	Black 1990	Black 1991	White 1989	White 1990	White 1991	Pacific Islander 1989	Pacific Islander 1990	Pacific Islander 1991
Grand Total, Women	271	357	394	271	357	394	271	357	394	271	357	394	271	357	394
Ethnic Subtotal			2 / .5	31 / 11.4	36 / 10.1	39 / 9.9	1 / .4	5 / 1.4	2 / .5	231 / 85.2	307 / 86.0	334 / 84.8	2 / .7	3 / .8	
Biochemistry				3 / 1.1	3 / .8	7 / 1.8		1 / .3	1 / .3	31 / 11.4	21 / 5.9	21 / 5.3	1 / .4		
Biology			1 / .3	2 / .7	1 / .3	1 / .3	1 / .4			40 / 14.8	43 / 12.0	50 / 12.7	1 / .4		
Chemistry				3 / 1.1	4 / 1.1	2 / .5				16 / 5.9	16 / 4.5	13 / 3.3			
Earth Sciences										8 / 3.0	9 / 2.5	16 / 4.1			
Applied Math/Statistics					2 / .6					6 / 2.2	2 / .6	4 / 1.0			
Mathematics						1 / .3				4 / 1.5	6 / 1.7	8 / 2.0			
Physics and Astronomy				2 / .7	2 / .6	3 / .8				11 / 4.1	7 / 2.0	10 / 2.5			
Behavioral Sciences			1 / .3	6 / 2.2	4 / 1.1	9 / 2.3		2 / .6		56 / 20.7	68 / 19.0	73 / 18.5			
Biomedical Sciences				1 / .4	1 / .3	2 / .5				25 / 9.2	25 / 7.0	28 / 7.1			
Computer Sciences				5 / 1.8	1 / .3	1 / .3				5 / 1.8	10 / 2.8	4 / 1.0			
Engineering			1 / .3	9 / 3.3	19 / 5.3	13 / 3.3		2 / .6	1 / .3	29 / 10.7	100 / 28.0	107 / 27.2		3 / .8	

136

TABLE D (continued)

Discipline	Mexican American			Puerto Rican			Cuban			Other Hispanic			Unspecified		
	1989	1990	1991	1989	1990	1991	1989	1990	1991	1989	1990	1991	1989	1990	1991
Grand Total, Women	271	357	394	271	357	394	271	357	394	271	357	394	271	357	394
Ethnic Subtotal	2 .7		5 1.3	2 .7					1 .3		2 .6	7 1.8	2 .7	4 1.1	4 1.0
Biochemistry	1 .4		1 .3												1 .3
Biology											1 .3	2 .5		2 .6	
Chemistry									1 .3						
Earth Sciences															
Applied Math/Statistics															
Mathematics												1 .3			
Physics and Astronomy															
Behavioral Sciences			2 .5	2 .7							1 .3	2 .5	1 .4		2 .5
Biomedical Sciences	1 .4												1 .4		
Computer Sciences															
Engineering			2 .5									2 .5	2 .6	2 .6	1 .3

* Includes biochemistry, biophysics, and molecular biology.
** Prior to 1991, this field included psychology, economics, and sociology. Because the disaggregation of behavioral sciences——into (1) anthropology, sociology, and linguistics; (2) economics, urban planning, and history of science; (3) political science, international relations, and geography; and (4) psychology——did not occur until 1991, a single category is used here.
SOURCE: Office of Scientific and Engineering Personnel.

137

TABLE E: NSF Minority Graduate Fellowship Program Applications by Women, by Ethnic Group, 1989-1991

Discipline	American Indian			Alaskan Native			Black			Pacific Islander		
	1989	1990	1991	1989	1990	1991	1989	1990	1991	1989	1990	1991
Grand Total, Women	398	436	644	398	436	644	398	436	644	398	436	644
Ethnic Subtotal	15 / 3.8	20 / 4.6	21 / 3.3		1 / .2	1 / .2	219 / 55.0	257 / 58.9	384 / 59.6	17 / 4.3	14 / 3.2	17 / 2.6
Biochemistry/ Biophysics	3 / .8	3 / .7	2 / .3				49 / 12.3	53 / 12.2	64 / 9.9	3 / .8	2 / .5	3 / .5
Biology	3 / .8	2 / .5	4 / .6			1 / .2	20 / 5.0	15 / 3.4	30 / 4.7	3 / .8	2 / .5	1 / .2
Chemistry/Earth Sciences		3 / .7	4 / .6				20 / 5.0	17 / 3.9	18 / 2.8	4 / 1.0	1 / .2	4 / .6
Physics/Math/ Astronomy		1 / .2	3 / .5				26 / 6.5	32 / 7.3	42 / 6.5	2 / .5	3 / .7	2 / .3
Behavioral Sciences	8 / 2.0	8 / 1.8	5 / .8		1 / .2		81 / 20.4	95 / 21.8	135 / 21.0	4 / 1.0	3 / .7	4 / .6
Engineering	1 / .3	3 / .7	3 / .5				23 / 5.8	45 / 10.3	95 / 14.8	1 / .3	3 / .7	3 / 5

Discipline	Mexican American			Puerto Rican			Cuban			Other Hispanic		
	1989	1990	1991	1989	1990	1991	1989	1990	1991	1989	1990	1991
Grand Total, Women	398	436	644	398	436	644	398	436	644	398	436	644
Ethnic Subtotal	41 / 10.3	56 / 12.8	71 / 11.0	54 / 13.6	42 / 9.6	73 / 11.3	3 / .8	8 / 1.8	22 / 3.4	49 / 12.3	38 / 8.7	55 / 8.5
Biochemistry/ Biophysics	6 / 1.5	10 / 2.3	11 / 1.7	19 / 4.8	13 / 3.0	18 / 2.8		1 / .2	2 / .3	9 / 2.3	8 / 1.8	10 / 1.6
Biology	6 / 1.5	11 / 2.5	9 / 1.4	4 / 1.0	5 / 1.1	5 / .8			4 / .6	7 / 1.8	5 / 1.1	7 / 1.1
Chemistry/Earth Sciences	1 / .3	1 / .2	6 / .9	5 / 1.3	9 / 2.1	10 / 1.6		1 / .2	1 / .2		3 / .7	2 / .3
Physics/Math/ Astronomy	4 / 1.0	2 / .5	6 / .9	2 / .5	1 / .2	7 / 1.1	2 / .5		2 / .3	3 / .8	4 / .9	2 / .3
Behavioral Sciences	14 / 3.5	25 / 5.7	31 / 4.8	20 / 5.0	8 / 1.8	23 / 3.6	1 / .3	3 / .7	7 / 1.1	25 / 6.3	8 / 1.8	23 / 3.6
Engineering	10 / 2.5	7 / 1.6	8 / 1.2	4 / 1.0	6 / 1.4	10 / 1.6		3 / .7	6 / .9	5 / 1.3	10 / 2.3	11 / 1.7

* Includes biochemistry, biophysics, and molecular biology.
**Prior to 1991, this field included psychology, economics, and sociology. Because the disaggregation of behavioral sciences——into (1) anthropology, sociology, and linguistics; (2) economics, urban planning, and history of science; (3) political science, international relations, and geography; and (4) psychology——did not occur until 1991, a single category is used here.

SOURCE: Office of Scientific and Engineering Personnel.

TABLE F: NSF Minority Graduate Fellowship Program Awards to Women, by Ethnic Group, 1989-1991

Discipline	American Indian			Alaskan Native			Black			Pacific Islander		
	1989	1990	1991	1989	1990	1991	1989	1990	1991	1989	1990	1991
Grand Total, Women	42	61	63	42	61	63	42	61	63	42	61	63
Ethnic Subtotal	2 / 4.8	4 / 6.6	4 / 6.3			1 / 1.6	17 / 40.5	25 / 41.0	27 / 42.9	1 / 2.4	4 / 6.6	4 / 6.3
Biochemistry/ Biophysics	1 / 2.4						3 / 7.1	5 / 8.2	3 / 4.8	1 / 2.4	1 / 1.6	
Biology							1 / 2.4		1 / 1.6		1 / 1.6	
Chemistry/Earth Sciences		1 / 1.6	1 / 1.6				3 / 7.1		1 / 1.6			1 / 1.6
Physics/Math/ Astronomy			2 / 3.2				1 / 2.4	3 / 4.9	3 / 4.8			
Behavioral Sciences		1 / 1.6	1 / 1.6			1 / 1.6	7 / 16.7	9 / 14.8	10 / 15.9		1 / 1.6	1 / 1.6
Engineering	1 / 2.4	2 / 3.3					2 / 4.8	8 / 13.1	9 / 14.3		1 / 1.6	2 / 3.2

140

Discipline	Mexican American			Puerto Rican			Cuban			Other Hispanic		
	1989	1990	1991	1989	1990	1991	1989	1990	1991	1989	1990	1991
Grand Total, Women	42	61	63	42	61	63	42	61	63	42	61	63
Ethnic Subtotal	6 / 14.3	8 / 13.1	7 / 11.1	4 / 9.5	6 / 9.8	5 / 7.9	1 / 2.4	4 / 6.6	4 / 6.3	11 / 26.2	9 / 14.8	12 / 19.0
Biochemistry/ Biophysics	1 / 2.4	3 / 4.9	2 / 3.2	2 / 4.8	1 / 1.6	1 / 1.6				1 / 2.4	1 / 1.6	
Biology	2 / 4.8	1 / 1.6	1 / 1.6						2 / 3.2	1 / 2.4	2 / 3.3	3 / 4.8
Chemistry/Earth Sciences				1 / 2.4	1 / 1.6			1 / 1.6				1 / 1.6
Physics/Math/ Astronomy			1 / 1.6							1 / 2.4		2 / 3.2
Behavioral Sciences	1 / 2.4	2 / 3.3	3 / 4.8	1 / 2.4	2 / 3.3	3 / 4.8	1 / 2.4	1 / 1.6		6 / 14.3	1 / 1.6	3 / 4.8
Engineering	2 / 4.8	2 / 3.3			2 / 3.3	1 / 1.6		2 / 3.3	2 / 3.2	2 / 4.8	5 / 8.2	3 / 4.8

* Includes biochemistry, biophysics, and molecular biology.
**Prior to 1991, this field included psychology, economics, and sociology. Because the disaggregation of behavioral sciences——into (1) anthropology, sociology, and linguistics; (2) economics, urban planning, and history of science; (3) political science, international relations, and geography; and (4) psychology——did not occur until 1991, a single category is used here.
SOURCE: Office of Scientific and Engineering Personnel.

141

TECHNICAL APPENDIX

The National Research Council (NRC) has a long and distinguished record of involvement in activities designed to increase the rate of participation of women in scientific and engineering careers. Recognition of the need to secure the fuller participation of women in the sciences and engineering and the role of NRC in achieving that end was one of the concerns in the 1972 establishment of the Commission on Human Resources, evidenced by the inclusion in its organizational plan of a committee to identify and work toward the solution of problems related to the education and employment of women in science and engineering. From 1973 until 1982, studies by the Committee on the Education and Employment of Women in Science and Engineering (CEEWISE) contributed significantly to an understanding of the issues involved.[8]

Since 1981 the Committee on Women's Employment and Related Social Issues (WERSI) within NRC's Commission on Behavioral and Social Science and Education (CBASSE) has reviewed, assessed, and encouraged research in the area of women's employment and brought research findings to bear on the policymaking process. Highlights of a meeting convened by

---

[8] Among the CEEWISE reports are *Women Scientists in Industry and Government: How Much Progress in the 1970's?* (1980), *Career Outcomes in a Matched Sample of Men and Women Ph.D.s: An Analytical Report* (1981), and *Climbing the Ladder: An Update on the Status of Doctoral Women Scientists and Engineers* (1983).

CBASSE in 1983 and chaired by Mary L. Good, current chair of the National Science Board, are given here:[9]

- In his introductory remarks, Robert White, president of the National Academy of Engineering, noted the importance of reexamining the situation of women in science and engineering, "since there have been significant changes in the proportion of women in the . . . profession over the past few years."
- Shirley Malcom, director of the Office of Opportunities at the American Association for the Advancement of Science, gave a keynote address on the main issue, emphasizing "We're better off than we've ever been, but we are not as well off as we ought to be. How do we get to the next step?"
- Edward A. Knapp, then director of the National Science Foundation, spoke of the difficulty "not with getting women in [science and engineering], but with helping women move up in their careers."
- James Hirsch, president of the Josiah Macy, Jr., Foundation, discussed the Foundation's support of programs to encourage women to enter scientific and engineering disciplines (beginning in the 1960s) and noted "that the reasons for career under-achievement [should] be given a high priority in future work."
- Jean Fetter, associate dean for graduate studies at Stanford University, focused on graduate enrollment, degree completion, and employment of women in science and engineering, suggesting that further study of their role in industry be examined.
- Gertrude Goldhaber, Brookhaven National Laboratory, "suggested

---

[9] Quotations appearing in this section were taken from an unpublished summary of the CBASSE planning meeting on women in science and engineering, September 9, 1983.

that future work in this area examine the conditions that enable women to contribute the most to science, . . . [recommending] the early encouragement of boys and girls in scientific activities, a review of hiring and tenure rules, and the continuation of the data base on women Ph.D.s in science."

- Lilli S. Hornig "recommended that a new committee [on women in science and engineering] develop a strong policy focus [and that] the Academy should also consider making public statements about these issues."

At the same meeting, other suggestions focused on players, policy issues, studies, structure, and priorities:

- There should be a "high-level statement on the part of the Academy regarding women in science and engineering concerns," continued attention to career advancement issues, and considera-tion of the impact of teacher preparation.
- Career advancement is a critical issue.
- The new committee "should focus on practical, policy-oriented, comparative research, coupled with dissemination to policy makers. . . . The committee and its funding should be organized around issues: access, advancement, effects of technological changes on women, and data maintenance and acquisition."
- Committee "membership should include representatives from the following categories: (1) senior people who make science policy decisions, (2) personnel directors from industry, (3) academe, (4) sophisticated social scientists, (5) research directors in institutions, (6) statisticians, (7) government, and (8) media."
- The committee should have two roles: "(a) research and infor-

mation and (b) advocacy. . . . The need for a center for research on women in science and engineering was underlined."

- Studies undertaken by the committee should deal with "access, attrition, and advancement."

The underrepresentation of women in science and engineering was identified as a problem that WERSI should address, and women from scientific and engineering professions were appointed to the committee and its working panels to undertake several relevant studies:

- *Women's Work, Men's Work: Sex Segregation on the Job* (1986) reviewed women's position in the labor market and documented both changes and stability in job segregation over the past two decades. That report indicates the negative consequences for women of continued sex segregation in the workplace; evaluates several explanations for continued job segregation; addresses the effectiveness of remedies that have been attempted at both the public and private level; and offers several policy recommendations to reduce segregation.
- *Sex Segregation in the Workplace: Trends, Explanations, and Remedies* (1988) further addresses various aspects of job segregation. These reports also identified some of the structural barriers to increasing the participation of girls and women in science and mathematics.
- *Computer Chips and Paper Clips: Technology and Women's Employment* (1986) was most concerned with these changes as they are transforming clerical work.
- *Pay Equity: Empirical Inquiries* (1989) contained the research of 11

146

small-scale studies on the wage determination process and comparable worth policies.

- *Work and Family: Policies for a Changing Workforce* (1991) synthesized and evaluated the research in three major areas: (1) the effects of different employer policies (e.g., scheduling policies, benefit policies, leave policies, etc.) on working families; (2) the effects of employees' family circumstances (e.g., dual-earner versus female-headed) and responsibilities (e.g., child care and elder care responsibilities) on their work availability, commitment, and performance; and (3) the factors that influence employers to adopt new, family-related policies (e.g., size, industry, economic conditions).

Subsequently, when NRC was reorganized in 1984 and activities of the Commission on Human Resources were transferred to the Office of Scientific and Engineering Personnel (OSEP), interest in this issue remained. OSEP was charged with the responsibility for activities of NRC that contribute to the more effective development and utilization of the nation's human resources, giving emphasis to national education and manpower utilization programs and needs.

Other units of NRC have also been interested in enhancing the progress of women in the sciences and engineering, however. For instance, at a three-day conference in 1984, the Transportation Research Board (TRB) examined the role of women to examine ways to meet the challenges associated with transportation education and training.[10] Among the

---

[10] See, for instance, Lillian C. Liburdi, *Education and Training Needs of Women in Transportation,* a paper prepared for the Transportation Research Board, 1984.

147

papers commissioned on this topic was "Education and Training Needs of
Women in Transportation" by Lillian C. Liburdi of the Port Authority of
New York and New Jersey. Beginning with a basic premise—

> in the competitive, deregulated, economic environment
> that transportation faces today, with the international
> pressures to continue achieving the type of growth
> experienced in the past, businesses must capitalize
> strategically on all of the resources at their disposal—

Dr. Liburdi stressed the need for greater research on the role of women in
the transportation field. She particularly noted the need for studies
"assessing the skills that contribute to a successful transportation career and
analyzing the career paths of successful male transportation managers in
order "to generate clues about career paths . . . useful for role modeling
and as a foundation for career planning for women in transportation"
(TRB, 1985). Citing discussion at the September 1983 meeting organized
by CBASSE, Dr. Liburdi noted that studies of career choice and
development patterns of men and women in science and engineering—
including family conflict, residual barriers resulting from sex bias, and
successful institutional policies to reduce barriers to participation by women
in science and engineering—should be the focus of research. In summary,
she presented seven initiatives that should be considered:

1. conduct a more comprehensive examination of the education and
    career paths of transportation and nontransportation industry
    executives to determine factors of success;
2. determine whether specific educational backgrounds are more
    likely to lead to success;
3. determine the number of women in the various transportation
    industries and the kinds of positions they hold;

4. determine why women continue to be underrepresented in managerial roles in transportation businesses;

5. identify barriers to career advancement for women;

6. evaluate the support structure for women, including mentoring and role models; and

7. determine feedback methods and opportunities for promotion and growth, including on-the-job and off-site training, to learn whether there are differences in the process for men and women and, if so, why.

Subsequently, TRB established its Task Force on Women's Involvement in TRB and in Transportation in the spring of 1986 to examine the extent to which such involvement might reasonably be expected; the extent and nature of barriers to involvement; and potential actions appropriate for consideration by TRB. That task force, jointly with the Task Force on Minorities' Involvement in TRB and Transportation, developed a questionnaire sent to the chairmen of its 150 standing committees to determine barriers to the involvement of women and minorities in TRB and to recommend mechanisms for eliminating such barriers. The task force also collected data on female members of TRB committees and staff, as well as data about women who are potential members of these committees; began to highlight sessions to TRB's annual meeting that would be of particular interest to women; and encouraged related organizations to support individuals interested in attending that meeting.

Along the same lines, the Government-University-Industry Research Roundtable (GUIRR) sponsored a symposium on nurturing science and engineering talent in 1986, noting that

Early socialization that limits opportunities for girls to

149

engage in activities that develop their scientific interest and competence is a key factor underlying women's under-representation in science and engineering. The remains of historic exclusion, now expressed in covert job discrim-ination, and conflicts between professional achievement and family responsibilities also affect women's career choices. The impact of these interrelated factors is evident in data showing that, in high school, fewer female than male students take advanced mathematics and science courses that are critical to technical careers.

GUIRR also noted that

Many observers perceive discriminatory practices——reflected in unemployment, underemployment, salaries, and rank and tenure——to be the most serious impediment to the goal of equality of opportunity for women in science and engineering education. The recent increases in the participation of women indicate that improvements in opportunity do result in improvements in participation.[11]

During the past eight years, NRC has employed women more extensively in advisory and evaluative roles. At the Water Science and Technology Board——a joint project of the Commission on Engineering and Technical Systems (CETS) and the Commission on Physical Sciences, Mathematics, and Applications (CPSMA)——Sheila David, senior program officer, has compiled a directory of female mathematicians, scientists, and engineers who are interested in serving on committees on the National Research Council. To begin the directory, women who had served on NRC committees from 1982-1984 were asked to provide biographical data (if they were interested in serving on future committees) and to suggest the names

---

[11] Government-University-Industry Research Roundtable, *Nurturing Science and Engineering Talent: A Discussion Paper* (Washington, D.C.: National Academy of Sciences, July 1987).

of other individuals who, although not members of the three institutions, could provide expertise to NRC. CETS and CPSMA staff use the directory when creating committee nomination packages and have been successful in forming committees whose membership includes 2-5 females. Data in this directory have now been computerized by OSEP staff, who use it primarily for compiling rosters for panels to evaluate applications to the fellowship and associateship programs that it administers for NSF.

Recent demographic changes indicate that NRC must now assume a more proactive role. At an OSEP-convened workshop in 1986, participants explored what is known about the causes of the observed underrepresentation and differential participation of women in science and engineering at all educational levels and about the patterns and causes of their differential career development relative to that of men.[12] One consequence of that workshop was a call to OSEP for advice and information in setting priorities for research and action programs. In response, OSEP was granted approval by the chairman of NRC to develop a program of research about the education and employment of women in the sciences and engineering. To that end, NRC provided funds in June 1988 for the convening of a Planning Group to Assess Initiatives for Increasing the Participation of Women in Scientific and Engineering Careers. The purpose of the Planning Group was to determine how the OSEP should configure its activities in order to best fulfill several objectives:

- sensitize key decision makers about the seriousness of the demographic problem;

---

[12] Linda S. Dix, ed., *WOMEN: Their Underrepresentation and Career Differentials in Science and Engineering,* proceedings of a workshop (Washington, D.C.: National Academy Press, 1987).

- illuminate the multifaceted issues that must be addressed in reaching solutions to that problem; and

- clarify and interpret the voluminous and complex research on the underrepresentation of women in science and engineering.[13]

To meet these objectives, the planning group surveyed the activities of other organizations and met with representatives of various federal agencies and other groups committed to increasing the participation of women in scientific and engineering careers. Because the gender imbalance in participation can be a subtle, difficult issue to address, many within the science and engineering communities believe that the issue has been resolved. It was pointed out at these meetings, however, that the issue has not disappeared. In addition, the Planning Group determined that an important strategy for ensuring an adequate supply of U.S. scientists and engineers to meet pressing national needs in an increasingly global marketplace would be to increase the representation of women from all racial and ethnic groups in scientific and engineering careers. The planning group's major recommendation, therefore, was the establishment within OSEP of a continuing Committee on Women in Science and Engineering, whose role would be to undertake activities designed to increase the participation of women in scientific and engineering careers. This recommendation meshed with NRC's mandate to OSEP to strive for

> enhanced strength as a resource and center of expertise
> . . . concerning the status of scientific and engineering
> manpower and the methodologies for assessing current
> and projected employment demand and personnel supply.

---

[13] The Research Council's provisions of funds in 1972 for a similar activity, the Conference on Women in Science and Engineering (June 11-12, 1973), had led to the establishment of CEEWISE.